Skeoch

*Our new life on a
Scottish hill farm*

Jennie Phillips

ALSO BY JENNIE PHILLIPS

A Bolt For Freedom

BY CONRAD PHILLIPS

Aiming True – An Autobiography

All books published by Any Subject Books
(www.anysubject.com)

ISBN
Paperback edition: 978-1-909392-12-0
EPub edition: 978-1-909392-13-7
PDF edition: 978-1-909392-14-4

Disclaimer: This is a work of non-fiction containing references
to real locations however all characters appearing in this work
are fictitious. Any resemblance to real persons, living or dead, is
purely coincidental. If, on reading this, you feel that there are any
inaccuracies in any of the salient details, please bring it to the
attention of the publishers immediately.

Cover image: © *Can Stock Photo Inc. / mrdoomits*

CONTENTS

FOREWORD

It was whilst writing this book that it came back to me just exactly how wonderful our time at Skeoch was. The totally unspoilt way of life was really a soul-nourishing experience and one that I would most certainly recommend to the young, foolish and brave. I feel very privileged to have had the experience which is something I shall treasure for the rest of my life.

I am very grateful to the lovely people there, that were so supportive to this pair of novices. To have found such unparalleled beauty was something to treasure and helped to build our strength and resilience against sometimes enormous odds. It equally gave our children a wonderfully early start on the appreciation of nature in all its forms.

We were at Skeoch for six years. It taught us a lot.

Thank you Scotland!

Jennie Phillips

February 2013

CHAPTER 1

Pulling the woolly balaclava hat off my head, I opened my eyes and peered over the bedclothes. A dim half-light, filtered through the window. The inside of the window was covered with ice that had made opaque, intricate patterns on the glass. I shivered. Somewhere a strange animal screeched. It was about 6.30am. Lying back on the pillow, I glanced around the room. Rather lurid, old-fashioned pink wallpaper decorated the walls and an elderly, tweedy pink carpet, the floor. There were no other concessions to luxury. I ached in every limb and the temperature in the room seemed well below freezing. I felt as though I had spent a night training with the S.A.S.

Slowly my disorientation disappeared and a little lurch of excitement touched my stomach as I remembered. Today was January 2nd 1972. Yesterday we had moved. Gone the luxury flat, the fitted carpets, the central heating, the hot water. Here, instead, we were in a little stone-built cottage, half way up a mountain - in Scotland, in the freezing cold. With a grate! In total isolation.

I lay for a while watching the still, recumbent figure of Conrad, my husband, all cosily cocooned in the folds of the eiderdown, his breath condensing on the air!

Too early to get up yet, too cold. Kate, our little daughter was still asleep in the corner of the room. I snuggled into the warmth of Conrad's back and let my mind drift back to the beginning.

This then, is the story about my husband, Conrad Phillips, a professional actor, my daughter Kate of two and a half, my 20 year old unemployed stepson Patrick and me - a pregnant wife, who all left their comfortable London house in Fulham and sallied forth into a world of self sufficiency and a new, clean life, away from care-worn London, city smoke and After Eight Mints. This is the story about Skeoch, a place that was to change my life completely.

I had met Conrad in the April of 1968, when I was working in a small commercial film production company as a P/A cum do-it-all, finally graduating to the dizzy heights of Casting Director/P.A./cum-do-it-all!

Arrow Productions was run by a charismatic, ebullient chap called Ken Davies and the company was situated about 83 stairs up, on the top floor of a magnificent building in Orme Court in London. There was no lift, so there was no alternative but to hammer up and down the stairs several times a day. The staircase itself was a most impressive thing, wide sweeping with long brass stair rods, that were kept highly polished by Dot our lovely Welsh cleaning lady.

We shared the building with the two comedy writers, Ray Galton and Alan Simpson and the comedians Frankie Howerd, Eric Sykes and Spike Milligan! It was a fun time, when BIBA was THE place to shop and all you needed was love, love, love.....

Conrad and I met through a Curry's commercial that Arrow Productions were making and, as the Casting Director, I liaised a fair amount on the telephone with him about fee and wardrobe etc.

It was to be a 60's commercial, based on a James Bond type theme, with Conrad dressed as 'James' and a gorgeous lady with long blonde hair, as his....assistant. (Also cast by me.) It was filmed in a disused warehouse, somewhere in Camden Town. Not the most exciting location to meet your future husband.

Through a series of minor disasters in the office, I missed the main part of the film shoot and finally met Conrad for the last five minutes, on the last day of filming.

We married ten weeks later in Chelsea Registry Office. You could call it a whirlwind romance I suppose.

Due to a divorce settlement, Conrad left his previous marriage with absolutely nothing. In spite of starring in a well known television series and working in many films and the theatre, we started our married life together penniless, in one room, off the Bayswater Road, with only a gas ring on the wooden floor and a single bed. That was two years previously and it was the Swinging Sixties and "all things were possible in this, the best of all possible worlds".....

As I became pregnant on our honeymoon, Conrad extended our single bed to accommodate my increasing girth. He was destined never to be without a screwdriver in his hand for almost the remainder of his life, poor soul!

We laughed a lot, loved a lot and lived on the princely sum of £10 a week, including rent and food - but the accommodation was becoming cramped with the arrival of

Kate, our daughter and even more so when Patrick, Conrad's 17-year-old son by his previous marriage, arrived on our doorstep one evening, with his small brown case and asked if he could come and live with us.

We got another room in the house, then another but the need to have something of our own, was becoming more important.

The place of our own finally became possible when Conrad landed a year's hard work on a BBC television series called The Newcomers. This enabled us to put down a deposit on a large Victorian semi-detached house in Fulham and enabled us equally, to get a mortgage. It was a big, impersonal house, but redeemed itself with some nice features. High moulded ceilings and a few pretty fireplaces and the traffic thundered by outside.

The whole of the downstairs of the 13 rooms, we let, to young, upwardly mobile people, retaining only the top self-contained flat for ourselves. Patrick moved in with us and had his own room on the floor below, with enough privacy for him to feel unrestricted but also safe in the knowledge that we were only upstairs if he needed us.

For a time it worked perfectly well and we were a happy, contained unit with our daughter Kate, who by then was 18 months old.

After I left Arrow Productions to have Kate, I did temporary work occasionally, as and when we needed an extra injection of cash. There was no shortage of freelance work in London then and it fitted in very well with our life style. When Conrad acted I stayed at home and looked after Kate. When he didn't, he stayed home and I worked. It was the modern marriage and Conrad was very happy to embrace another crack at fatherhood. However, after a year, we started to feel a bit jaded. 'Fash Fulham' was losing it's charm and I was appalled that every time I took Kate for a walk on Parsons Green, we could only walk about 2 yards before coming into contact with dog mess and litter. There were not the doggie laws that currently apply. After a shower of rain on a hot Sunday afternoon, the smell that rose up from the pavements was disgusting.

The local infant school, that Kate in due course would attend, was jammed to overflowing, with little faces crowding at the bars of the playground. It was a large Dickensian red brick building, with very little space for the children to skip and play around. The need for some fresh, clean air, green grass and a bush or two, was becoming more apparent. I just couldn't picture Katie in that school. We were feeling claustrophobic.

One day Conrad came home from filming, looking quite excited.

"You know Jen, I've been working out our finances and I reckon, if we re-adjusted our life style, I could get enough money together to buy us a cottage, somewhere in the country".

It seemed a wonderful idea and it became the dominating theme in all our conversations. Although it certainly seemed a good possibility, strangely enough I also felt rather apprehensive. I was brought up, for a lot of my childhood, in some of the more remote parts of the country. Small villages in Devon and Suffolk and I had a creeping fear that we might find ourselves in some outlandish area, with no street lights, outside loos and one village shop. I wasn't sure I liked the idea. I mean, it all looks very pretty on a post card.....

We talked things over and finally agreed that it would

be best if we looked for a week-end cottage that we could go to for holidays, perhaps gradually improve and 'do up'. Somewhere, we thought, near the coast. Dorset seemed ideal.

The idea gradually took shape and so we went, one dripping Saturday, down to Dorset. The rain lashed the windscreen in a steady deluge. The sky was grim and laden with plenty more to come. Not a promising start. Wearily we squelched to various estate agents and looked at various properties but any that we really liked were totally out of our price range. The bracing wind and gun metal grey sea, only served to depress us further. Our gloom was contagious and in desperation, we headed back for London, abandoning the idea of Dorset. It is amazing to think in retrospect, that if the day had been sunny, our whole outlook might have changed, we might never have moved up to Scotland and we would not have so radically changed our lives.

Our life settled down again in it's irregular pattern. Having a husband who is an actor, can be many things - but one thing it isn't - and that's dull. We continued to alternate between being 'rich' and 'poor'.

I have to state right now that we never had any intention of buying a farm. Least of all in South West Scotland. It wasn't one of those nice, rational decisions that we all think we will make, when the time comes. It was totally irrational and made absolutely no sense at all - but we made it just the same!

The whole event was rather precipitated by my mother and sister who, dissatisfied with their lives in Bedfordshire, where they had been living for some time, left and finally took the High Road, in an attempt to find some security, peace and proper tranquillity. They were now happily ensconced in a grey granite house, built to keep the Scottish winds at bay, right on a wild bit of coast, at a place called Gatehouse-of-Fleet, in Kirkcudbrightshire, on the Solway.

Through the end of their garden, was a path that led down to a totally private stretch of beach, that continued for miles. It was a piece of heaven.

On a week-end telephone call from my mum, despondently we told her that we had been looking for a country cottage, without luck.

"You sound as if you could both do with a break" she said. "We've settled in beautifully now. Why don't you come up and stay with us for a few days?"

It seemed like a good idea and a week or so later, we headed up North.

We left Fulham on a grey, cold, drizzling Autumn morning. The car was filled with flasks of coffee, sandwiches, maps and hastily written instructions from my mum, none of which seemed to make sense. Katie was bundled in with the travel rug and toys, juice and emergency nappies and off we set, emerging some 6 hours later into the quaint little town of Gatehouse-of-Fleet.

Our stay was really what we needed. A complete break away from London and leisurely walks along the secluded beach with Kate, exploring rock pools and making sand castles.

We looked at various cottages in the immediate area, more out of curiosity than serious intent. One little house in the nearby village of Castle Douglas, some four miles away, was offered for sale with a guide price of £1,800. We had a look round. It needed a fair bit doing to it but it was very solidly constructed and quite charming. In a moment of madness we put in a sealed offer for it, almost as a joke - and crossed our fingers. It was too good to be true and we were not surprised to be pipped to the post. I think it went for £2,000.

Reason told us that anyway Scotland was too far. Logic told us it was totally out of the question. We had a lovely few days in the countryside and decided to forget the idea of a country property for the time being.

Haste Ye Back the roadway sign read as we thundered back over the border and headed for London. Was I imagining a lump in my throat when I read it? Probably not - it turned out to be morning sickness!

A few weeks later my mother phoned again.

"Jen are you still looking?" she enquired.

I told her that we had given up the idea for the time being but that we had discovered that I was pregnant again. She was delighted and after all the grandmotherly things being said, the conversation veered round to the fact that she had heard of a farm for sale, some 40 miles from her. It wasn't a very big farm. She thought it was up a track; possibly up a small mountain. She was a little vague, but it was very reasonably priced, she had been told. Why didn't we at least go and look at it?

Conrad and I both laughed. A farm. Ridiculous. We knew nothing whatever about farming - out of the question. So much for reason and logic though because a few weeks later saw us north of the border again. I don't think either of us seriously thought anything would materialise from the trip. However....

We studied the map. Turn right off the A75. Fine. Through the tiny village of Shawhead. That must have been what we had just driven through and blinked. OK. Unbelievably pretty countryside and BIG skies. Winding country lanes. No street lights here, I noted! Hmm. Follow the road, turn left at the village Post Office (one village shop, I also noted). Turn right. A small brook burbled and rushed over stones, beneath a charming 'Brig-a-Doon' stone bridge, covered with the silvery grey of lichen. The weather was fantastic; the clear Autumn light illuminating the breathtaking views in fine detail. Lovely.

Finally Conrad pulled the car over.

"It's got to be around here somewhere" he said. "Let me have a look at the map".

We both peered at it.

"That has to be it - up there" he said, pointing vaguely to a pot holed, overgrown track, that disappeared into a mass of brambles and undergrowth.

"Are you sure?" I queried. "It looks as though it leads nowhere".

"There's one way to find out" he said. Sure enough, a battered sign, the paint peeling off, simply said - SKEOCH.

We followed the long rutted lane for some distance, upwards, passing slowly a tiny gypsy caravan, parked happily in a clearing, where the track widened a little. It's little chimney belching out clouds of wood smoke. A savage black Alsatian dog, leashed nearby, jumped and barked ferociously. We were starting to feel a bit apprehensive. There must be some mistake. We had obviously taken a wrong turning. Apart from the caravan, all we could see were fields and hills. Then, over to the left, the great purple mound of Skeoch Mount, looming large and forbidding. We carried on slowly up the rutted lane and finally pulled into a large concrete courtyard, surrounded by cattle byres and a Dutch Barn.

Nestling in a hollow, further on, sat a small, squat, stone-built cottage. We could see the chimney needed re-pointing immediately, as the smoke came out not only at the top but at various other points en route. We stopped the car and stepped out. We were half way up a mountain!

We wandered to the cottage door, trying to take everything in. After Fulham, the space seemed amazing. Not a building in sight - just acres and acres of fields, hills and trees and the air was so clean and fresh.

"Well, I must say," Conrad said. "I thought your Mother said it was a small farm! This looks bloody enormous. I mean - look at it".

I had to agree with him. There certainly seemed a lot of it but one didn't really know then, where this farm ended and someone else's land started. I couldn't wait to explore. The backdrop of Skeoch Mount gave everything a slightly theatrical air. It could have been a film set.

The front door opened.

"You've arrived then". Mrs Moody, the owner, stepped out to greet us, offering her hand.

She was dressed in an old grey tweed coat that had seen better days, tied around the middle with a length of bailer twine, stout lisle stockings and thick brogue shoes. She ushered us into the house.

Catherine Moody was an ample lady, with thick, white hair, cropped mannishly and gripped securely in a no-nonsense fashion, on the side. Her face was generous and canny, with more than a touch of humour around her mouth. Eyeing us quizzically, she gave us a very old-fashioned look when she discovered, during the course of conversation, that Conrad was an actor. You could almost read her thought processes....."An Englishman, a townie AND an actor. Very airy, fairy dear. Hardly farming material....."

However, the kettle was put on for a pot of tea, while we were shown round the cottage.

Comfortable was not one of the words that would spring to mind in describing it. The accommodation was extremely basic and comprised of a very small, unmodernised kitchen, with well worn lino flooring in a tempting shade of cow pat brown. By the side of this there was a narrow, unlit corridor that had a tiny loo and bathroom at the end, with flooring to match the kitchen. The sitting room was dark wood panelled with a very unattractive shiny brown tiled fireplace and grate as the sole feature, apart that is, from the post war metal, single glazed, casement window that looked out on to the back garden and view beyond. But what a view; what a superb, uncluttered, unspoilt, uninterrupted view. It was magic!

Through the sitting room was a little square hall, one bedroom to the left (the one in pink) and one straight ahead. A further storage room led off this, with old metal french doors leading to the garden. Tacked on to the back of the cottage was a rather broken 'conservatory' with odd broken panes of glass and peeling green paint. A sliding door in the hall revealed a small flight of attic stairs, at the top of which were two rooms with sloping ceilings and tiny velux windows. That was it. The walls in the kitchen were running with condensation and the fire in the grate coughed cheerlessly, throwing out hardly any heat but a good deal of smoke. Talk about Cold Comfort Farm! Yet there was nothing of major proportions that was wrong with the place. Nothing that some new decoration, cosy lighting and a few creature comforts wouldn't put right, we felt sure. Fortunately Conrad and I both have the ability to see through the dross, to the nugget of gold and are not easily put off by cosmetic imperfections! It was a stout little house in need of a lot of TLC but perfectly adequate for our family needs. At this stage you have to bear in mind that we still didn't really have a firm idea of WHAT we were going to do anyway.

We looked around the outside with Mrs Moody. Two stone pig sties sat in a small field to the left of the house. Across the wide courtyard, ran a series of whitewashed cattle byres in an L formation, with stable doors and shutters. A vast Dutch Barn, perched on a concrete base, at the edge of the courtyard, looked out over a patchwork of hills and fields. Not another building was anywhere in sight. There seemed so much of everything. So much to take in and after rambling over various acres, we came to a picturesque wood, with a lazy stream meandering through. It was enchanting. Mrs Moody caught the expression on our faces and smiled.

"It's beautiful" we murmured.

"Aye - it's two and a half acres in all" she said and after a pause "included in the sale".

I looked at Conrad - our own forest!

We finally returned to the yard in a rather euphoric state. The red and gold November setting sun was illuminating a beautiful rowan tree in full berry, on the corner of the courtyard. It looked as if it were on fire.

Before the afternoon was completely through, we found ourselves striking a deal with Mrs Moody and promptly drove into Dumfries, the nearest town, to Thompson, Roddick and Laurie - Auctioneers and Estate Agents, to sign the Deed. Just like that. The whole thing had a dream-like quality to it but this was no dream; this was committal. There was no way out now - we were becoming apprentice owners of a fifty acre hill farm in South West Scotland!

The real enormity of what we had done hit us a few days later, when some of the euphoria had left us and we were back in London. This was no 'week-end cottage' - this was 'change your life' time. What on earth were we doing?

CHAPTER 2

Once back in London, the paperwork for the purchase of Skeoch Farm went through very quickly. We advertised and found suitable tenants for our top floor flat in Fulham and were on the road and heading for the A75 early on January 1st 1972. New Years Day/Hogmanay...Hooray!

The successful letting of our London house enabled us to launch the whole Scottish project. We discussed in depth what such an enormous change of lifestyle would mean and Conrad felt reasonably confident that he could 'commute', as it were, from Scotland, as and when any acting work turned up.

We had traded in our little Renault 4 car for a white Renault van with a larger capacity, in which we packed everything we needed to start our new life. We left the furniture in Fulham and Mrs Moody had assured us there would be a double bed a table and chairs and a settee and a few other bits and pieces, so there was no need for a full scale furniture van.

So, here we were - it was January - and I was pregnant. The weather was minus EVERYTHING and each morning found me picking my way across the kitchen floor, trying to avoid the freezing water, which had run down the walls and window over night and collected in a large pool near the kitchen sink. So the first hazard of the day, was just getting to fill the kettle!

The damp permeated everywhere. Clothes never felt completely dry and apart from one open fire in the sitting room, there was no way to dry anything. We had no mains water, no mains drainage (never did find out what happened to all that!) and no central heating. We did have electricity though, although that was a bit 'iffy' on occasions. So we had a series of paraffin lamps with glass funnels.

Looking back, I don't know how we survived that first winter. Perhaps sheer enthusiasm for our new lifestyle got us through. It was the only time in my life that I went to bed in a hat, a vest or two, cardigan and nightie. I was sure my unborn child would suffer from hypothermia before it was born. Later of course, we installed central heating but in the meantime I thought longingly of my warm London flat, with copious amounts of hot water, warm rooms, fitted carpets and dry clothes. Then I would look out of the window and hear the total silence and feel the vast openness and peace, the all pervading timelessness and tranquillity. There was really no contest.

Work was coming in for Conrad at a steady pace, during this time. Before we left London, he had completed a first season with the English National Opera Company, playing a non-singing part, in fact the lead part, of Pasha Selim in Mozart's 'Seraglio'. We made the move to Scotland during his six week break, before the second season started in February. Along with the 'Seraglio' , he was also narrating for Stravinsky's 'Oedipus'.

It was extremely demanding work and not really the thing to have a split focus on. As the time drew nearer for him to go back to London, more and more time was spent on organising things at Skeoch, to make my situation a little less vulnerable. He was, after all, leaving a pregnant wife (who couldn't drive at that stage), a small daughter and a twenty year old son, who was not madly practical, up a mountain in the middle of winter. So, Conrad decided to buy a gun. The advantages were supposedly two-fold. It would be useful to bag the odd pheasant and rabbit if we ever needed it and to provide some small comfort that any intruder would get more than they bargained for. That was the theory anyway, although I doubted that anyone would get past the slavering jaws of the black Alsatian down the lane easily.

However, the .22 was bought and the tin cans were place at the bottom of the garden. I was taught to load and fire and re-load and fire and load and fire ....

Perhaps the fact that Conrad was a gunner in the Royal Navy during the war had something to do with it. I must confess to feeling terrified at first, convinced that I would probably hit the cat, or at the very least, hit something living! I had a lot of near misses with the cans but eventually felt reasonably confident by the time he was due to commence work back in London.

As it turned out, my target practice was very nearly put to the test.

Conrad was now back in London, having finally departed from Skeoch in the most appalling blizzard of the century, to slide nearly seven hours down the motorway. The strain was horrific, mile after mile of thick, whirling snow, the windscreen wipers barely able to cope with it and gale force winds hampering every inch of the way.

Patrick had also gone away for a few days to be with friends. I had put the usual full day in, had locked up and gone to bed quite early. Although only around four months pregnant, I was feeling more tired than usual and the baby had discovered the use of it's feet at last. I lay there trying to encourage sleep but couldn't seem to settle. I turned the light back on, read an ancient magazine and battled my way through two chapters of a rather tired historical novel and still felt no nearer sleep at 2.00a.m. when I finally put the light out. Settling down in the dark, I willed myself to find the 'white light' in the head and relax. Suddenly, I heard the light go on in the upstairs hall. Click! I lay in silence, terrified. Even my stomach stopped thumping. What seemed an age went by. In reality it was probably no more than a few minutes. There was no further sound from upstairs. I was sure I had locked all the doors. Could anyone have got in without me knowing? Carefully I groped for the gun (which was kept on hand by the side of the bed), my heart pounding. I prayed Kate wouldn't wake up as I tip toed past her bedroom door. Standing at the bottom of the attic stairs, I peered up. Yes, the attic light really was on. We had no furniture or carpet in the upstairs at this stage and the wooden stairs creaked under my weight. Up I went, the .22 neatly balanced across my stomach, voluminous tent-like nightie billowing at the sides and a warm baggy cardigan pulled across my front.

"O.K" I muttered, kicking open an attic door in true Western manner. "I've got you covered. Come on out before I let you have it". My heart was in my mouth. Nothing. I peered round the door. Still nothing. No black moving shadows. No heavy breathing, just nothing. Feeling foolish, I went over to the other door and did the same. The house was as quiet as the grave. I am sure that if anyone HAD been up there, they would have given themselves up immediately, at the apparition I made. Far more scary really. A wild-eyed pregnant lady with a .22

balanced across her stomach. A cross between Annie Oakley and Lorna Doon rolled into one!

It wasn't until I finally turned the light off upstairs, that I realised I hadn't even loaded the gun. Needless to say, I spent the rest of the night and the rest of the week, come to think of it, with my bedroom light on.

When Conrad returned from London for a few days, I told him what had happened.

"I suppose you think there's a ghost then" he said, smiling. "Complete nonsense Jen. it's probably just a faulty switch, or a mouse might have run up the wall and knocked it on". (Five feet up?). It's old fashioned. I'll change the light switch".

And so he did and that might have been the end of it, if we hadn't had friends come to stay about 6 months later, for a long week-end. Casually, on the Sunday morning, one of them asked what had been wrong in the middle of the night because the attic hall light went on about two-ish!

"Oh" I muttered rather lamely and staring right at Conrad "it's only a faulty switch, or a mouse might have run up the wall and knocked it on". They exchanged glances between themselves and said nothing. However, it never happened again - perhaps it was a mouse after all!

Later, when most of the unpacking and settling in process had happened, life started to fall into some sort of routine. Wednesday was usually the day that we went shopping into Dumfries, which was some 10 miles away.

It was a strange little town; a mixture of extremely pretty and extremely plain. The buildings were mostly no-nonsense granite in various shades of grey or red. Practical and thorough, with no frills or 'fancy stuff'. Built to last and withstand the buffeting Scottish winters. Conversely, the river Nith ran through the centre of the town, with pretty bridges, weirs and swans. This was Burns country, with statues and follies. Turn a corner and you could be in an ancient, cobbled street. Turn another and it could be quite grim. Even so, it was well served for shops and all amenities.

Now, this particular Wednesday morning, Conrad came into the kitchen, looking thoughtful.

"You know Jen, this kitchen is very small and the corridor next to it serves no purpose at all really, does it? I mean, it's completely dark and pokey. Do you think we really need it?" I agreed with him that I didn't see much point to it at all. he disappeared and came back a few minutes later, carrying various crow bars and a sledge hammer.

"What time do you want to go into Dumfries?" he asked.

"I was thinking in about half an hour" I replied.

"Right, well that's just about enough time then, I reckon" he laughed. "Let's go to it". With that, he took a mighty swing with the sledge hammer at the corridor wall.

Half an hour later, amidst dust, debris and chaos, we stood looking proudly at my now large kitchen. The corridor was gone, apart from a few upright posts (which were NOT load bearing) which we demolished later in the day. That was the start!

We got through the worst of that long winter. We survived the two feet of snow, the sleet and gales and although heavy frosts still festooned the trees and bushes with a delicate web of white, the spring bulbs imperviously pushed through, bringing the first bright daubs of colour. The vast mound of Skeoch Mount, behind the cottage, began to soften. The bruised purple and grey giving way to dappled lemons and bright greens and browns. The clouds, passing over it's face, in ever changing patterns, giving a kaleidoscope of shifting colour and light.

There was so much delight in those early weeks, the bright yellow Forsythia and daffodils, a pretty ornamental cherry with red leaves, blue pools of grape hyacinths. By the side of the house was a wild patch, filled with small fir trees, which in early spring, became completely carpeted in snow drops. The children would later christen it Fairyland. By the gate was a blousy japonica, resplendent in fiery orange flowers. Each month brought more pleasures. Solomon's Seal, bluebells, hypericum and rowan trees, drooling fire, their red berries blazing and dancing, caught in the first rays of an Autumn sun. Pheasants and rabbits were in abundance and our first entry into one of the outside cattle byres had us gazing ecstatically at a large white barn owl, that stared back at us somewhat belligerently.

Along with a few vital bits and pieces from London, we also brought with us our only cat. A large black Persian affair, bought a year previously from the Fulham Road veterinary 'Poor Box' as a very tatty but appealing kitten with huge yellow eyes, skinny legs and a matted coat. Now he was a fine sleek cat and very 'street wise'. He had diced with death frequently, with the London cab drivers. Zebedee, as he was called, immediately set about staking the joint out and lugged a fair-sized dead rabbit up the garden path, within a few days, to proudly show me. No pheasant was safe from this feline gangster either, and the goldfish in the deep stone pond in the garden hid in the reeds while his black, hairy paw combed the water, hoping for a quick snack. During that first winter, he would often be seen sitting on the ice in the middle of the pond, watching the fish swimming safely beneath him. He was doubtless hoping to hook one out. Rats were thankfully becoming fewer too, all being neatly dispatched along with some rather too adventurous moles, their little black forms casually left about the place, like so many discarded gloves.

By the end of April, I had only another three months to go before 'baby day' and my stomach was really quite large. It was still too cold to wear a skirt and I spent most of my time in some comfortably large black trousers. The zip in front unhappily no longer worked and so I strung a piece of elastic from one side of the waistband to the other, held together (I am ashamed to say) by a safety pin. Although it was a bit draughty, with a long jumper on they served their purpose. It was in this gear, with a towel over my head, having just had a bath (and my old flip flops on), that I spotted Zebedee through the sitting room window. In his jaws dangled a young rabbit that squeaked and kicked like mad. I dashed into the garden, aimed a stone at the cat, missed and started in hot pursuit after him. He fled down the garden path and turned sharp right, towards some old cow sheds, the rabbit still dangling wildly from his mouth. Just by the cow sheds I nearly caught up with him, when suddenly he leapt across a large stretch of what looked like new spring grass. Not to be outdone, I leapt after him but of course landed squarely in the middle. Then everything happened at once. Immediately I found myself up to my knees in the most foul smelling green slosh. The 'new spring grass' was in fact a large, very well rotted mound of ancient cow manure, on which the grass had happily grown. The flip flops bubbled a few times and disappeared from sight, the safety pin in my trousers, unused to so much exertion, chose that particular moment to undo and they floated gracefully to my knees, leaving a large expanse of heavily pregnant stomach exposed. The towel swiftly unwound from my head, leaving me stranded there, trouserless, shrouded in towel and cow manure, like some phantom of the midden. Then the phone rang back at the house. All this took about 30 seconds but by the time I had gathered my smelly self together, the cat, of course, had disappeared.

It is strange the things one has to adjust to, moving from a City to the country. One of them was the incredible quietness. We had become so used to the constant roar of London traffic churning past our house in Fulham all day and night, that the first few weeks at Skeoch were almost unnerving. Silence, peace, tranquillity, with only the weird night calls of some strange nocturnal animals. The first time I heard a vixen call, terrified the life out of me. There was a whole new set of sounds to get used to, as well as a whole new way of life.

Most of our time in the early months was spent on re-decorating, whitewashing the outside of the cottage and generally trying to do some urgent repairs to the chimney stack, which required Conrad tying himself to the roof and balancing a step ladder and bucket of cement at the same time, to try and re-point it. There was plenty of repair work to be done to the outbuildings and barns too. With so much to do the days flew by and we ended up most evenings, absolutely bushed, falling into bed in unfashionably early hours. My progressive pregnancy was beginning to wear me down a bit, especially as little Katie was a 24-hour-a-day job alone. At two and a half she was an active, healthy child, with an insatiable curiosity - and a lot of space to try it out on! From the confines of a flat in Fulham, to the unlimited space of Skeoch, she blossomed. There was so much to explore. So much hide and seek. Always an easy-going child, she took to the country life like a duck to water and, as my stomach got bigger, she began to look forward to the arrival of the new baby. Would it be staying with us always? Was it going to be sharing her bedroom?

Patrick on the other hand, found it all rather a shock. He began to show signs of boredom with the isolated lifestyle. Quite understandably from his point of view. As a young man he needed his own kind. The kind that preferred Pink Floyd, Led Zeppelin, Cream and lots of Jimmie Hendrix.

Work in that part of Scotland was just about non-existent, unless he wanted to be a farm labourer but his interests were not in agriculture. Whilst in London, he had worked as an Assistant Stage Manager on tour with Conrad for a theatre production and later he had a job in a record company, which suited him down to the ground. This part of Scotland was very wild and where we were in particular, quite isolated. Social amenities were thin on the ground. We bought him a little motor scooter, which gave him some independent mobility and he would occasionally go for a jaunt to 'swinging' Crocketford, where there was a rather basic bar, with some 5 locals (all over 60) and situated some 7 miles from us, or occasionally he would go as far as Dumfries - but it wasn't a happening time for him. It was rather a tall order to suppose he would find a life there. Where were the girls? Not up a mountain in Shawhead that was for sure.

CHAPTER 3

The long rutted cart track that ran from our farm was grossly overgrown with blackthorn bushes, brambles and hawthorn. At the bottom, it branched off to the right and continued down, eventually leading to another, much larger farm - The Scour. The occupants of this were our nearest neighbours, Jim Kissock and his daughter Doreen. Their family had farmed the rough hill land for generations, with cattle and sheep. Doreen did most of the lambing every spring, single handed with the aid of Pat their border collie dog. She also had a full time job in the local motor tax office in Dumfries. She was in her mid-twenties with light brown hair and merry, dark brown eyes; she had a great sense of humour and quickly became a special friend to me.

Half way down our part of the track was the tiny caravan we had noticed on our first trip. It was approximately 14 foot long and 5ft 6inches wide, with small stained glass windows and a tiny, smoking chimney. The proud owners of this establishment were Paddy and Millie. She was a small, bird-like woman, with dark unruly hair and about 5 teeth sparsely dotted about. She used to hawk the rags in a battered old pram around Dumfries, until she became Paddy's common law wife and then she hawked the rags around everywhere. Hawking was still an active pursuit among some of the poorer folk there in the early 70's. Basically, Millie would beg for old, discarded items of clothing from house to house - anything no longer useful. All were stuffed into her pram as she went round. Then she would try and sell to her various 'outlets'. The Rag & Bone man, the paper mill etc., for a very few pence. This supplemented her income. She was a very rare soul and was incredibly kind hearted.

By the side of the caravan, in a small portable run, was the large, black, ferocious Alsatian dog, (called, of all things, King,) which I confess frightened me to death. It barked incessantly and bared its fangs at the drop of a hat. The run was littered with gnawed bones of all descriptions and I was never entirely sure if they were not really the last earthly remains of some poor, unsuspecting traveller who had lost his way!

On wash day, which was generally about every three to four months, when the weather was fine, Millie would fill the great copper boiler that stood outside the caravan with buckets of water from the spring that bubbled between the dyke and stone boundary nearby. Once full, she lit a small fire underneath and plunged the linen, of all colours, in together. Grey, red, blue, more grey, Paddy's long johns, sheets and bloomers, scarves, odd socks and tea towels. Once in the boiler, there they stayed, poked at occasionally with a stick and an odd lump of cooking soap tossed in for good measure. It took all day and much refuelling of the smoky little fire, to get the water hot. When she was satisfied that the contents were 'done', she would hoik them out, rinse them in the spring, and hang them on all the bramble and blackthorn bushes around the caravan to dry.

One was never surprised to see more 'holey' than righteous undergarments festooning the lower branches of the rowan trees, in various shades of grey. There they stayed, hung out to dry until it rained or the wind blew them off, or just taken off an obliging tree when necessity demanded.

Paddy was an altogether different kettle of fish. He was a ditcher and dyker and did any other odd farm work he could cadge. A bantam cock of a man and an ex-fly weight boxer, having done some sea-side booths in his time. He was a great chatterer but a veritable 'devil' when the whisky was in him, which happened periodically about every four weeks, when his benefit money came through (commonly called, rather aptly - 'the brew'). Otherwise a drop of Meths or an odd bottle of Brasso seemed to do equally well. When sober he was niceness itself - but beware the demon drink!

We hadn't been at Skeoch very long, when I decided to call down and see Paddy. I had been told that he would get a large sack of potatoes for me, from Jim's farm. It was a blissfully sunny afternoon in June, when I made my way down to the caravan and timidly knocked at the door. King did his usual snarl in greeting and I knocked again, louder. Amid much cursing I was finally rewarded by the caravan door being flung open, revealing a very drunk Paddy. His eyes were also glazed over (the right one puffed up and puce) and he stared straight past my left ear; thickly he smiled and said "Hello hen, c'mon in ". I started to demure and tried to explain about the potatoes but King was roaring away behind me and the only answer was to go in. The smell inside that hot little bothy was indescribable. Stale sweat, stale tobacco smoke, mixed with unwashed bedding, a liberal sprinkling of sour whisky and a smouldering coal fire. To add to the general congeniality of the atmosphere, there were split peas and lentils. Not in a pot or a dish but everywhere. Floor, bed, chairs and cupboard. On Paddy's clothes, in his hair... and at the far end, a militant Millie, sitting up in her tiny bed (jammed widthways across the caravan), her eyes also glazed and in her hand, ready to take aim again, another whole bag of butter beans. She relaxed when she saw me come in.

"What are ye wantin' hen?" She gave me one of her toothless grins.

"Yon auld sod's been at it again". She jerked a thumb at glassy eyed Paddy. Judging from her lop-sided smile, with her left eye puffing up to a nice rich mauve, I rather gathered she had had a drop herself. Nervously I explained about the sack of potatoes, wishing with all my heart that I had never bothered but it was too late to back out now. Paddy industriously poured a vast quantity of neat whisky into a dusty denture mug and thrust it into my hands. I started to refuse but caught the look in his one good eye and thought better of it. I took a tiny sip and smiled, frantically trying to think of a way of refusing the rest and escape from the split peas and heat.

"You know the trouble wi' you English?" Paddy pushed his whisky sopped face into mine - "Ya canna' bloody drink - now down it hen, fer God's sake - it's good fer ye - and yer wain!"

I took a gulp. Millie visibly cheered. "Tha's it hen - gi' her a drop more Paddy, yer mean auld bugger". The nicotine stained hand grabbed the bottle, the yellow teeth smiled. Peace was being restored.

An hour later, I tottered out, impervious to King's gnashing teeth. I even managed a "Good Boy" as I passed, groping my way up the lane in the fresh, clean air.

After about 3 weeks, we had given up all hope of the potatoes, when a knock at the door produced a somewhat reticent Paddy. The faint traces of the black eye were still visible but gone were the split peas and belligerence. Touching his cap, he came into the kitchen.

" 'Scuse me missus but was it YOU wantin' some tatties, 'cos I was telt someone came down to see me a wee while ago an' I canna' quite just remember who it was". I told him I had paid him a visit but didn't want to push for them, so I had decided to leave it at that.

The next morning, on our doorstep, was a huge sack of potatoes and three large turnips! I made a mental note never to visit on a "Brew Day" in the future.

We first encountered our postman the day after we moved into Skeoch and he delivered our mail in a rather haphazard fashion, in a bright red Royal Mail van. Sometimes he arrived at 11.00 a.m. and at others 1.30 p.m. or later. You could never be sure and I suspect it largely depended on how many cups of tea and biscuits or how many drams were offered to him on his way up, as to what time we received the post. Either way, it was for us, the main event of the day for the most part, except when he delivered bills, which seemed to occur with alarming regularity. It was on one of his expeditions to us that we started a hen-laying venture that very nearly took us over.

"What you need" he said one morning "are some hens. You've plenty of space here and a few hens would be a good thing. You could always eat them when they got too old for laying".

We readily agreed. Apart from Zebedee we had no other animals. A farm was not a farm without hens and it would be fun to go out in the mornings and collect brown eggs for the table. I saw it all in my mind's eye!

"I've got this friend, whose got some to get rid of, cheap mind. I'll see what he says".

We were happy with the idea and off he went. I put it from my mind, until a few days later when he screeched up the track into the yard. I went out to meet him. Muffled squawks were coming from the back of the post van and with a flourish, he opened the door. Out rushed the most motley collection of hens you have ever seen, along with letters, parcels, feathers and lunch boxes and large polythene bottles full of petrol. The remainder of the letters in the back of the van, bore more "messages" on the outside of the envelopes, than on the inside. When the totally hysterical bunch had been rounded up and calmed down, we counted about a dozen or so hens, complete with cockerel. Brown, white, black - you name them, we had them! As if this was not enough, there were also some half a dozen bantams, also complete with cock, one rather twittish guinea fowl and two Muscovy ducks, who eyed us with complete mistrust.

We surveyed the collection around us, rather bewilderedly.

"Just gi' 'em a handful of grain once in a while and let them free range" was Andy's sage advice. He gave a chuckle, took £5 and with a cloud of exhaust, lurched back down the lane. That was just the beginning.

The hens quickly adapted themselves to their new environment, roaming freely around the farm and being shut into a spare cattle byre at night. Their favourite place of all though, seemed to be the long, narrow window box, which was placed outside my kitchen window. I had always intended to plant bright, sunny nasturtiums or geraniums in it but had never found the time. Just as well really as Hettie, Brownie, Goldie and Ruby would, each day, carefully wander down to their sunniest spot, to sit and cackle. Their brown and white fluffy bottoms perched in a row on the window box for me to gaze on while I did the washing up. Like some elderly grannies get-together, there they would stay for an hour or two each afternoon, squawking and clucking. No amount of shooing did any good whatsoever, as they would disdainfully shuffle off for ten minutes, only to re-appear when I was safely the other side of the window. However, after finding most of our young lettuce pecked to bits and various other tender garden produce reduced to shreds, we decided a compound was necessary. Conrad fenced in a large piece of land attached to the cow byre and there they were all forced to stay. All was fine, except at feeding time, when the two Muscovy ducks waded in and within a trice, shovelled the grain down, leaving very little for the hens. The more food I put in, the more "Bert" and "Bertha" ate, their flat bills acting like earth movers. The only solution of course, was to shift the handsome pair out of the compound. We had a family conference and decided that since our wood had a good natural pond and plenty of shelter, it would be wonderful for them to go free - and perhaps breed, in a totally natural environment. So, armed with large sacks, we finally cornered them, complaining indignantly, and took them down to the woods, some six acres away. There we set them free by the pond and waited. They appeared to be enchanted (if a duck can be). They waded, spluttered and basked - and Con and I congratulated ourselves on a job well done.

The next day I wandered into the compound to feed the chickens. There, waddling up to the front of the queue, as usual, were Bert and Bertha. Again we cornered them and took them to the woods. Again they flew back to the hen compound. In the end we ruefully clipped their wings to prevent them flying back. Presented with this 'fait accompli' they finally adapted to their woodland setting and quite often, when we went for a ramble, they were to be seen roaming around. They never did breed though and finally, after some years, they disappeared. Whether they just moved on (there were no confines or barriers), or whether they were supper for a hungry fox, we never knew - but they were a very colourful pair of characters.

The bantams were incredibly... abundant. Each season brought more and more. When bantam hens are broody, they are ferocious and it is very hard to get too near when they are sitting on a clutch of eggs. Several of ours just disappeared into the undergrowth, only to re-appear some weeks later with anything up to twenty little fluffy balls cheeping along behind! Multiply this half a dozen times and you have a serious bantam breeding problem. Each batch of course, produced more young cockerels. Now Conrad and I are not 'natural born killers' by nature but twelve bantam cocks was just too much, even for our kind hearts. Some just had to go. We were also not very adept at the wringing of necks. Work in the theatre, TV and films had not encouraged that part of our repertoire. What to do, what to do?

With solemn intent, Conrad strode purposefully into the hen compound. A man with a mission. First of all he shut away all the hens and one lucky bantam cock (called Austin - named after Conrad's double in the William Tell TV series of the 50's). Over his shoulder the .22 gun - High Noon. The rest were given a generous handful of corn - blissfully unaware that this was their last supper!

Their final resting place was, in fact, Fairyland, now about four foot high in nettles. Not a place Katie would go roaming in too quickly - we hoped!

CHAPTER 4

Although we had been living up in Scotland now for about seven months, I still found it difficult to always understand the local dialect - especially if there were several people talking together. Sitting in the little bus from Shawhead to Dumfries was an amazing experience. With all the local people travelling to the market, the conversation was abundant. The bus served as a good meeting place for everyone to catch up on their gossip. Each stop of the bus produced a Mrs McCleod, or a Mrs Drummond and the conversation would start anew. There was, of course, a great deal of curiosity about the English family, moving up from London and being in the "theatrics" and all, and occasionally a question would be kindly aimed in my direction, two thirds of which I didn't understand and, after making them repeat it again, was still no nearer understanding, I usually muttered something incoherent and hoped I had said the right thing. They probably thought I was simple, too!

The nearest hospital to us was about ten miles away in Dumfries and I had been going once a week to the pre-natal classes. Our baby's birth was rapidly drawing nearer - the doctors had reckoned on or about the 28th July. All was going very well my end, although my diet seemed to consist mainly of raw cooking apples, Rice Crispies and iron pills!

At about lunch time on the 27th July, I felt a twinge in my stomach. Not a big twinge - but nevertheless - a twinge. The twinges lasted all day, still not getting any stronger and I was not overly concerned. Conrad, on the other hand, was much more concerned. I think he had pictures float through his mind of having to deliver the child single handed at the farm, or in the back of the Renault van en route. By 10.30 p.m. I was tired and went to bed. The twinges were the same, not painful but a little quicker and I felt convinced they could go on for days in exactly the same way, without any ultimate panic. Suddenly, however, Conrad sat up in bed. "I think that's it," he said.

"That's what?" I enquired, feeling quite nice and drowsy.

"I think you ought to get dressed. I've been timing the twinges and you've had two in the last 15 minutes".

Conrad was present at the birth of our first daughter Kate, so he wasn't exactly a novice at what to expect. Still, he was taking no chances. Waking a sleepy Kate, I got dressed, grabbed my case and got in the van. Patrick was away for a few days, so we locked up and bumped and rocked our way down the track to arrive at Dumfries hospital - quite quickly really!

The birth of our Katie in St. Mary's hospital in Paddington, three years before, had been a very unsatisfactory experience for all of us. Not something I wished to repeat in a hurry. So it was with just a bit of apprehension that I viewed my labour in a little Scottish country hospital. However, my fears were immediately dismissed by the appearance of a brilliant lady doctor. My contractions lasted all night and all the following day, and by 7.00 p.m. the following evening a doctor told Conrad to go home and come back after supper. Both he and Kate had been hanging round the hospital off and on since about 9.00a.m. and the new baby didn't appear to be in any hurry to make its entry into the world.

"She has HOURS to go yet," one official said. So, off they went, planning to come back and see me at about 9.00p.m.

As is so often the case with these things, the moment Con and Kate left, things took another turn and speeded up. Suddenly it was all systems go and labour started in earnest. My nice lady doctor took total charge of the situation and had me well monitored. She was caring, considerate and compassionate and worked me through all stages of labour, as a team. In the very last stages, just before the baby's head appeared, she told me to stop pushing for a moment and listen. I stopped and heard the amazing muffled cry of the child within me, before it emerged. There was nothing wrong with its lungs obviously! At 8.09 p.m. I gave birth. At 8.40 p.m. utterly exhausted but exhilarated, I rang Conrad from the delivery room.

"You've got another daughter" was about all I could manage. He and Kate had just had supper and were at my bedside in what appeared to be minutes. They stared in amazement at our tiny, purple-faced, black-haired, very grumpy daughter. We called her Sarah and she weighed in at just five pounds thirteen ounces. I had fulfilled Conrad's dearest wish - and that was to have two daughters!

Sarah spent much of her time during the next three

days, yelling. On the second day, a passing mother popped her head round the door of my room.

"Hello hen" (this much I could understand). She peered at Sarah and shouted above the noise "ach what a boony lot of hair he has. Yer wain - does he greet much?" she smiled questioningly. I asked her to repeat the question (trying to buy time).

"Yer wain" she yelled - "does he greet all the time?"

"No, no" I stammered. "My - er wain doesn't greet at all thank you".

The kind lady wandered off into the corridor, looking more than ever perplexed. Finally I buttonholed a passing nurse and asked her what it meant. She looked at me with some amusement as Sarah reached yet another crescendo -

"It means, does your baby cry, Mrs Phillips," she smiled and bustled off.

On day three Conrad came to the hospital as usual but looking quite crestfallen. He sat down by the bed.

"You look as though you've lost a pound and found a sixpence" I said. "What's up?"

"My agent's been on the phone," he said flatly. "I've been asked to do another job. A very good job as it turns out but it means I would have to be away for a couple of weeks if I took it". He looked very doubtful and concerned.

"But that's wonderful," I said, giving his hand a reassuring squeeze.

"Don't worry about it - everything will be fine I'm sure

and we do need the money, let's face it. What date would you have to go?"

He looked very uncertain.

"Tomorrow" he said.

"Tomorrow - but I've only been here 3 days," I burbled "and I'm not due home for another four".

"I know, love," he said. "I can't possibly go now". He gave me a hug.

"I'll ring my agent back and say I won't take it. I just wanted to talk it over with you".

We were between a rock and a hard place. We needed the money, it was a good job.

"Look," I said, "I feel fine, really. There's no reason why I should hang on here anyway. I would feel happier at home. Just keep your fingers crossed that they will let me out of here!"

We had further chats and I promised to ring him as soon as I had seen the doctor.

After he left, I felt rather depressed. Not really the deprivation of four further days in hospital perhaps ...but... by the time afternoon tea was brought round I was seriously snivelling. The nurse stopped and smiled.

"Never mind, my dear," she patted my hand. "We all get a little depressed afterwards - it's all very natural - just reaction. After all, giving birth is a tremendous thing for anyone to do".

"It's not that," I blubbered, a germ of an idea forming

in my mind.

"It's my little daughter. She's missing me so much. She's only three. I have to go home, she's really pining". I squeezed out another tear. The nurse looked a little worried.

"I'll have a word with the doctor and see what he can do," she said and bustled down the ward. About an hour later the doctor arrived on his rounds. He eyed my critically.

"What's this now Mrs Phillips, you've a wee daughter pining at home I'm told, is that correct?" I nodded feebly and tried to look pathetic.

"Well, well. We can't have that can we?" he smiled kindly.

"I see no reason why you can't go home tomorrow. You're in good order but you must take it easy and get plenty of rest. We don't want you back again because you've overdone it. Have you someone to look after you when you return home?"

I smiled sweetly. "Oh yes, doctor," I lied, "thank you so much. My little girl will be so happy". So will my husband, I thought.

The next day I was discharged. Of course my 'wee daughter' was not missing me at all. She had the complete undivided attention of Conrad and Patrick as well, who had come home for a few days. A beaming Conrad picked us up from the hospital and slowly we drove back up the rutted lane to home sweet home. We spent the morning doing the last minute packing for Con's trip and he finally, reluctantly, left for London that same afternoon.

The realisation that all was not well was when I went to fill the kettle in the kitchen. A muddy trickle of brackish water spluttered from the cold tap and stopped. There had appeared no problem earlier in the day. Alarm bells were ringing in my head. We had a very quaint system of getting water, which basically meant that the spring travelled down the side of Skeoch mount to our lower fields, where it gathered in a shallow well. From there it was pumped uphill again into an old iron reservoir (in the same field) and then went by gravity feed, into the house. The trouble arose when there was a drought situation. Water was then being pumped (it was a very ANCIENT pump) out of the well, faster than it was filling, which in turn meant that the pump eventually sucked air, got airlocked and 'voila' no water in the reservoir. this was a drought situation, being the end of July. With nappies to wash (no such luxury as disposables then) and bottles to sterilise, this was posing one hell of a problem. The only thing I could do was to turn off the pump (which meant a climb up onto the roof of a cattle byre via an old ladder!) so that the well would eventually fill up. No bathing and certainly no loo flushing! In the meantime, we did have a very tiny catchment well near the side of the house that provided a small bucketful of khaki coloured water about every five hours. This is what I used. It was while this jolly little lot was happening that I also discovered that a bat had taken up residence in the sitting room, living, it appeared, quite happily behind the wood panelling. I slept for three nights with the sitting room windows wide open all night and finally succeeded in getting it to move on.

Two days after I returned from the hospital, Millie made one of her rare visits to the house. She wanted to see the new 'wain'. I took her through to the bedroom where Sarah lay in her carry cot on our bed. Millie bent over, tears glistening in her eyes. She cooed and muttered to the baby and thrusting her hands into her pinny, brought out a

10p piece which she pushed gently into Sarah's tiny clenched fist. The baby's fragile fingers closed over it and she held it tight. Millie was ecstatic. She looked up at me, grinning her toothless grin and wiping her eyes at the same time.

"Th'a's good luck hen" she beamed. "If she can hold the silver now - she'll never be wi'out it. It's an auld Scottish custom". Well pleased, I took her into the kitchen and made her a cup of tea. I knew Millie was desperately hard up but this was her gift to the baby. She turned out quite right too. A month later, I had to go to London to see Conrad, who was still working well. I took the train from Dumfries with both the children and we got seats quite near the buffet car. Just about everyone passing backwards and forwards, pushed money into Sarah's hands for good luck. By the time we reached London we had amassed about £3. 50 in 10p pieces!

One of the many delights that first summer was the abundance of strawberries. My mother had a fine strawberry bed at Gatehouse of Fleet and dug up a large pile of suckers, put them on the bus 40 miles away, where we collected them from the bus driver at the road end at Shawhead. We planted them in the spring. The summer that Sarah was born, we got a fantastic crop. It is a wonder I didn't get the size of a house, as I collected pale gold cream in a large jug from one of our neighbouring farmers, who had a dairy herd, and lavishly poured it over bowlfuls of sun warmed strawberries - almost every day, with much help of course, from little Katie.

The garden ran mostly to the back of the house and went down in a series of terraces, ending at the bottom in a rather shabby heap of soil. In time Conrad would remove it all and turn the unsightly pile into a beautiful lawn and flowerbed, which we massed with sweet peas.

One year we had a visitor from England come to stay with us for a few days bed and breakfast. he was, among other things, a judge for a rather impressive horticultural society. He was tremendously taken with our show of sweet peas that year, some of the stems having 5 - 7 flowers on each and sporting a dazzling array of colour and perfume. He told us we would walk away with First Prize at any of the shows he judged and asked what product we used to achieve such incredible blooms. The answer was quite simple. Just loads and loads of well rotted cow manure. it was the perfect answer to getting rid of the large grassy 'midden' that I had got stuck in, the first year chasing Zebedee.

Sarah by now had started growing at a real pace. The mass of black hair that she had been born with had disappeared to be replaced by a downy head of light brown hair, and the five and half pound bundle was turning into a delightful, chubby, healthy baby. She had a wonderful sense of humour at a very early age and was christened 'the giggler'. Kate was enchanted with her new sister, rarely letting her out of her sight; she wanted to help bathe her, change her nappies, give her a bottle and I must confess, even at that age of three years, she was of enormous help to me, taking the role of elder sister very seriously indeed.

Patrick too, adored both the girls and spent ages making them laugh and making 'nests' out of his sleeping bag for them. For some reason this was a very popular game and quite often I would come into Patrick's room and find them all giggling away in the 'nest' with Patrick on all fours, fooling about, pretending to be a lion.

Even so, his desire for his own independence was becoming more and more of an issue. What had started as a great 'drop out and back to nature' scene for him, was losing its charm as the isolation gradually crept in.

Eventually he said he wanted to return to Devon, where he still had a lot of friends of his own age, that he went to school with and where he could hopefully get a job. Reluctantly, in the end, we had to let him go. It would have been wrong to have tried to contain him, when his heart was elsewhere.

So much had been packed into our first half year at Skeoch, reaching a pinnacle at the end of July with the birth of Sarah. We were a happy, completely self-contained, loving family. However, the toll on Conrad was beginning to mount up. He was constantly driving between Scotland and England to work, hammering along motorways to grab a quick week-end with us, before charging back down the motorway on Sunday afternoon for work, either at the Coliseum or theatre, or wherever. This was great inasmuch as his career was happening and he was in demand but between to-ing and fro-ing, he was repairing, maintaining and improving the new homestead. Unbeknown to us all, the bill was coming in rapidly.

Disaster struck one November morning, when we had been to Castle Douglas for a change, to shop. Mid-way crossing the road, Conrad went dizzy, felt strange and looked absolutely terrible. Somehow he drove us home. He went straight away to bed and I called the doctor immediately. The local doc, arrived within a very short space of time and gave Conrad a thorough examination but he could find nothing identifiable. Conrad couldn't stand up - the room spun round. He felt nauseous and weak; his eyes were completely out of focus. The doctor was mystified. There would obviously have to be some tests done. The illness had taken us all by surprise. In a state of shock, it took some time to register just how serious it appeared to be. Con, normally fit and healthy, always bounced back from any bouts of flu or colds but this was no dose of flu. He was going down hill fast and he looked awful.

Blood and urine samples were taken, along with other intensive tests. Weeks went by. The doctors could find nothing conclusive. They couldn't be sure. Sure of what though? We were in the dark completely.

I was a bit like a Zombie during that time. Mechanically I cooked, washed, attended to the girls - trying to keep them as quiet as possible. Waiting. Waiting. The doctor called up every few days to see if there was any improvement - there wasn't. One side of Conrad's face seemed paralysed, his eyes were completely crossed and he had terrible vertigo every time he moved - even in bed.

In December he was moved to hospital for observation. It was Christmas. Sarah was now five months old, Katie three and a half.

'T'is the season to be jolly'....... I decorated the tree. I wrapped the presents and placed them underneath, wondering if half of them would ever be opened. I remember being on the floor of the sitting room with both children on Christmas morning, trying very hard to be cheerful and jolly. Katie kept asking why Daddy wasn't there. I explained that he was in hospital, that he would be home soon and we would have Christmas all over again together when he got back. We made paper chains to welcome him home! I wandered into the kitchen and looked out of the window. It was a true traditional scene. We had a white Christmas and it was very beautiful. The snow blanketed the yard. Trees shivered, their icing sugar branches illuminated brightly against the dark sky. All was hushed and solemn. I filled the kettle - with tears as well I think.

Thank goodness my mum and sister came over to stay

for the next few days. They gave me enormous support, emotionally and physically and helped, to some extent, to take my mind off things. It was a desperate time and the only antidote was work. Work as much as possible and try not to think. Easier said than done, we all know. The children, young as they were, seemed to sense the seriousness of the situation and were wonderful, providing me with a focus and much joy and love and I thanked God that I had them.

Finally Conrad was let out of hospital and brought home. We were still none the wiser about his illness and he was not improved. He lay in bed, frustrated and sick. He couldn't read and the radio, children and I, were his only consolation.

January 23rd dawned, cold and grey. It was my 28th birthday. Doctor Sloan called up to see Conrad. After he had examined him, he beckoned me into the sitting room.

"There is no easy way to tell you this" he said " but it's just possible that Conrad has a brain tumour". He waited for this to register for a moment, then continued.

"If it is, it could be that it is inoperable. We can't find any obvious tumour". He stopped.

"You do realise what I am saying?" he said gently.

"If it turns out to be a buried tumour, there is nothing we can do. He could be dead in three months, maybe sooner. I think it best that you don't tell him this at present. There is no point in taking him into hospital where he will only fret. However, there is one more test that we can do, that should show conclusively if it is a buried tumour and I have arranged for him to go to hospital in Glasgow and have this test done". With that and a few more kind words, he left.

Con went into Glasgow Infirmary and had the test, which involved having air injected into his brain while he was conscious and tracking it's course around his brain, while he was strapped in a revolving chair. If they were to lose sight of the air bubble, that could be where the tumour could be.

The result - definitely NO tumour. He returned home. He had a massive headache for days afterwards and felt ghastly. Now of course, people are scanned in a tunnel but we are talking about over 30 years ago and modern technology was not at its best. A virus was talked of. Disseminated sclerosis was talked of. Meanwhile gradually, he started to turn the corner. his vision started to improve marginally. His balance got a little better. He started seeing only two of everything instead of three and he could stand for a few moments without keeling over. He wore dark glasses for some considerable time, while his eyes were re-adjusting. It was a long, slow haul.

One thing that emerged from all this was that Conrad was told by the doctors, that he would probably never be able to act again. We were advised to sell our London house and somehow utilise the 50 or so acres of land on the farm.

After much discussion and as soon as Conrad was physically able to cope once more, we decided to sell the house in Fulham and consolidate. Through a friend in London, we quickly found what we thought was a buyer. We were assured that the house was just what they were looking for and we couldn't believe our luck that a sale could come about so quickly. They wanted us to proceed with all possible haste and empty the property of everything. We shot down to London, having reluctantly given notice to all our tenants and worked like demons, dismantling all that was necessary and emptying the entire 13 rooms of their contents. We were exhausted at the end of the week but very relieved and happy, when we headed back up the motorway again, the van stacked to the gunwales with items that we could not dispose of and that might come in handy on the farm. We were due to exchange contracts 2 days later. After the long drive back up to Scotland, Conrad, who was still very fragile, was extremely tired. His physical abilities had been stretched to their limits so soon after the illness and for a time I feared he would have a relapse.

Impatiently we stayed by the 'phone for the next 2 days, waiting to hear from our Solicitor that exchange of contracts had taken place. Imagine our horror then, when the 'phone finally rang and we were told that our buyers had decided to pull out at the last minute. They had found another property that they were going to proceed with instead. No apologies, nothing. By now, having emptied the house of furniture, fittings and tenants, we were in the unhappy state of no income on top.

Angry, frustrated and deeply upset, we had no option but to return the property to the market with a London estate agent - and wait. Now, was the worst possible time to try and sell. By 1973 the wonderful property boom in London had run its course and we were left with a falling market, a mortgage, falling house prices and no buyers. We waited, and waited...and waited. It took another eight months to finally agree a sale and when the buyer was finally found, the price had dropped to an all-time low. In the end we just wanted to get rid of it and try to start a new life.

While Con was in hospital, he asked many questions of some of the other patients, several of whom were farmers.

He asked about cattle and land and general husbandry. We knew absolutely nothing about farming and hadn't a clue what we were going to do with the place. Somehow we would have to utilise the land as profitably as possible. We had very little money to float any venture and we were apprehensive. Now we had no alternative. With Conrad unable to act - somehow we had to make a living.

CHAPTER 5

We were to embark on a new adventure. A complete change of life. I was unaware then of what a dramatic change it would be. I think I had the divine idiocy of youth at the time. Somehow we would survive, be happy, adjust. I really didn't have much time to think about myself and when I did it was always totally incomparable to my previous life in London. What did I miss up there on my mini mountain? The cinema? Occasionally. Sainsbury's? A bit. English pubs? Definitely. Especially those that we frequented in our early life together. 'The Queen's Elm', 'The King's Head' etc.

But those were fleeting moments, some nostalgic thoughts for a second or two that flit in and out again. We had too many other things to think about. Adapting, adapting all the time, to a life that I had only either read about in a magazine or heard about through friends of friends. I mean, farming hadn't really been on our London agenda at any time.

Now that we had definitely decided to try our hand at farming, having no practical experience at all didn't seem to really daunt us too much. Good husbandry seemed to be, for the most part, a matter of common sense - but of course there is always the element of risk. Always the 'unforeseen' which catapults you into an arena where you have very little, or sometimes no, control at all. Times when all the odds appear to be stacked against you. We wouldn't escape of course - who does? Conrad used to say - "Given average luck we might just pull it off, we might just be able to succeed" and so we hoped really for "average luck"!

So, it was one of those incredibly bitter cold mornings, with about a mere 20 degrees of frost. The sky was black and the wind was picking up speed, cutting through everything like a glacial sword. The once soft, bracken-laden hills looked dark and forbidding and the top of Skeoch Mount was obscured by a large, depressing cloud. The shrill ring of the telephone pierced the morning gloom. The slightly guttural voice of Jim Kissock on the other end of the 'phone, breathing heavily, announced... "I have this cow. She pulled her calf bed out a few days ago an' died. The calf's still alive - 'though I doubt it'll make it. She's had no sup for the past few days and I canna get her to drink. Do you want it?"

Did we WANT it? With ill concealed excitement, we leapt into our wellies, grabbed a sack and jumped into the van, bumping our way down to our neighbouring farmer. Jim was a solid well built man, who suffered permanent high blood pressure.

"She's in here," he announced, as we drew up. He lead us into a very draughty cow pen, that smelt dank and fetid. It was cold in the extreme. Crouched in one corner, shivering and feeble, lay a tiny brown and white Hereford calf. She gazed at us with sickly eyes, her frail body wracked with spasms every few minutes as she lay on the sparse covering of wet straw.

"Aye, there she is," said Jim. "I doubt she'll make it to the back end of noon," he said cheerily.

"Has she had anything to drink at all?" queried Conrad.

"Och, nought but a wee sup, about a half o' pint, a couple of days ago".

We looked at each other. What could we do? We knew absolutely nothing about rearing calves.

"We'll take her," said Conrad, looking at me. He could see I was already entranced by her.

"We'll just have to find out what to do".

Con got the sack out of the back of the van and he and Jim slid her neatly into it. She didn't seem to mind at all and showed no sign of a struggle. I think she was so ill, really she didn't care at all. Certainly her weight couldn't have been more than half a hundredweight and Con carried her easily to the van. She looked very pathetic with just her brown and white face sticking out of the top of the sack. We turned the van, declined "a wee dram" with Jim, anxious to get her back home. Jim re-appeared with a few pounds of powdered calf milk in a bag.

"It's worth a try I suppose" he said. "Mind you - I've no had the vet to her. There's really no point. If she lives - she lives - but if she dies well that's that. I'll tell you one thing though, if she lives, she'll cost you £25 - and if she dies - forget it". He waived us off and slowly we started the frost-bound ascent back to Skeoch.

There was great excitement and activity when we reached the other end. The first thing we had to do was find a clean, dry bed for her. Katie raced around grabbing loose straw that had been sitting in the bottom of the Dutch Barn. Soon the calf had a bed fit for a heifer! It was about 2 foot deep all over. Gently we unravelled her from the sack and plopped her onto her bedding. She just sat there, where we had left her, still shivering like mad. We then realised that we had no calf teat with which to try and feed her. Necessity the mother of invention came to the rescue in the shape of one of Sarah's discarded bottle teats, with the hole enlarged. It was a bit dainty for a calf's mouth but better than nothing. We stuck it onto the end of an old lemonade bottle, half filled with warm water and glucose, forcing it into her freezing mouth. She showed absolutely no desire to suck. It took us about 2 hours to get about 3 tablespoons down her but at least it was something and we began to hope. Every hour we went out and persevered again. By early afternoon she had painfully struggled to her spindly legs. Her stomach was shrunken in so much that she looked positively skeletal and emaciated. The tail half of her was caked in diarrhoea and she smelt appallingly. A few minutes later, armed with old cloths and a bucket of warm disinfectant soapy water, we set about cleaning her up. Gently we washed and dried her rear end. She made no movement and seemed content enough just to stand there patiently until we had finished. Now at least, we could see her little tail. Since her shivering still continued, we decided to sew her into her own "body stocking". Somewhere in our cupboards, I had kept an old Mexican type striped blanket, which I found and cut into an oblong. We threaded bailer twine through a series of holes in the side then wound it along the length of her and tied it securely along the top with more bailer twine. By that evening there seemed a definite improvement. She was walking about and the ague seemed to have disappeared. Con and I went in to see her after an excited Katie was finally packed off to bed.

"You know Jennie, she would be a lot better if she could

only eat something. She won't touch the hay," Con remarked. Finally, after various discussion, we decided that a bowl of porridge liberally sprinkled with glucose and cooled calf milk powder and water, and as a final ingredient, a generous slug of whisky. Clutching the bowl of warm 'gue' we made our way back to the byre. There was no way we could spoon it down her but she decided she wasn't adverse to the idea of licking one's fingers, so the only answer was to laden our fingers with the porridge and let her lick it off. In this charming fashion, we managed to get a fair bit down her and ourselves as well but satisfied we had done our best, we left her with a water bottle tucked around her and went back into the freezing night and house and thankfully to bed.

The farmer who owned a parcel of land on Skeoch Mount and whose land bordered our own, came up to count his sheep and cattle. In all weathers he came. Rain, snow, hail, gales - never missing a day. This morning he saw us scurrying about with a bottle and buckets and stopped by our gate.

"'Sa grand mornin'," he smiled, looking at the white brushed landscape. He spat and looked around with a twinkling eye.

"Aye - a grand mornin'".

"Come and see what we've got," we said, like a pair of school children. Over he came to the byre where "Lu-Lu", as she was now christened, was now walking around and really looking quite chic in her Mexican outfit. He peered in and stood looking with an incredulous expression on his face.

"Is that a calf in yon byre?" he queried. We nodded.

"I've never 'afore seen one dressed in its pyjamas," he quipped. We told him the story of how we acquired her.

"Do you think she'll live?" I asked.

"Ah wouldna' care to say," he replied non-committally. "Has she taken anything?"

"Well she's had some glucose and water and about half a bowl of cooked porridge," we said.

The old man fell into a spluttering chuckle.

"By God man, if that's no killed her, it's no harmed her either," he grinned and, with his dog, turned on his heel and wandered up the hill. We could still hear him laughing half way up.

Paddy came up and had a look at her and later, so did Jim. All left in hysterics, thinking no doubt that only the English would be mad enough to put their cattle in blankets.....'Did they give them a hot water bottle too? Ha-ha'! Damned right we did and an odd nip of whisky to keep out the cold.

From then on Lu-lu improved daily. Within the next 24 hours she "took to the bottle" with a vengeance and made rapid progress. We were delighted with her. She was the start. She was our herd!

By now we had central heating installed in the house and basked in the luxury of warm rooms. The condensation on the walls and windows that had plagued our lives at last stopped, and although as yet I still had no washing machine (all laundry being done in the bath) - it was marginally better than Millie's arrangement - I did at least have radiators on which I could festoon our undies. These were the days pre disposable nappies and the constant need for clean dry ones was paramount.

Finally, triumphantly, Conrad came home from a working trip down South, with a top loading washing machine, which he had spotted at an auction in Leeds and put in the van. This we filled with hot water from the kitchen tap, the clothes were put in, the motor turned on and that was it. It did not automatically rinse. The clothes had to be taken out and the whole process gone through again. In the end I usually rinsed everything by hand - but it did wash!

I had to make a trip to London for a few days and Conrad was left looking after the children and farm. On my return they came to meet me at Dumfries station and there was much excitement. Something was afoot.

"Just wait and see what awaits you when you get home," was all I got. All of them giggling conspiratorially. Katie had obviously been sworn to secrecy and was finding it very difficult to keep.

I went into the sitting room and there - what a sight. While I had been away, Conrad had completely demolished the old, brown tiled fireplace and chipped away all the surface plaster above to reveal a HUGE open hearth. You could actually walk into it and see the stars at the end of the chimney. A vast stone lintel ran along the top for support. It was absolutely amazing. Carefully he had carted all the rubble away. It made a wonderful focal point to the room and before very long he built little stone seats on either side of the hearth. The huge lintel and chimney breast stones were carefully re-pointed and scrubbed and the stonework was finally given several coats of linseed oil and turpentine, which brought up their natural glowing colours.

In time I designed a large wrought iron fire basket, which our local blacksmith (who was actually a lady) had made beautifully for us. She was a very lively lady and liked the design of the basket very much. We got chatting and I told her that I was a 'part time' artist. Since we had arrived at Skeoch, I hadn't had very much time at all to indulge in painting (except on walls) but I had produced a few landscapes, which I was reasonably pleased with. Out of the conversation she commissioned me to paint "a grand wild stallion, rearing atop of a mountain" as her daughter was passionate about horses. It seemed rather an ambitious project but I was pleased to accept and in due course completed the painting, which she loved and it also earned me a little extra cash. In time I was able to paint more and more. The beautiful countryside around us and the sometimes ethereal light flooding the hills and woods provided limitless inspiration. I sold quite a number of paintings too and saw myself progressing well in an area I had always loved.

The dreariest job for any farmer or landowner must surely be the repairing of fences. In our case it was no exception. The dry stone walls that bordered every field were extremely practical. They used up the rocks and stones that seemed to appear on the surface of the fields like a crop of acne every few months. If made well, they lasted for centuries and made very efficient wind breaks. Dry stone walling is an art. Not one, I hasten to add, that I am any good at. You can't just balance one rock on top of another and hope for the best. There is a proper symmetry. Small stones must be stacked and packed in between the larger ones and the walls themselves must be a good two feet thick if they are to withstand the winds and wildlife. Frequently we would pace over the fields to find twenty or so of our neighbour's sheep happily eating our grass. They would see us coming and make off over another wall, usually knocking ten feet or so out of it in their exit. We would curse and fume. It was a time-consuming, back-breaking pastime and a constant chore.

On the odd occasion we would wake up in the morning to discover sheep in the garden too, contentedly munching all our spring cabbage and vegetables. Skeins of peas hanging from their mouths like spaghetti! Their feet trampling our tender plants and flower beds. We would fling on our boots and yell blue murder at them. They in turn, would just crash through another wall, leaving a trail of debris in their wake.

Wild farm cats were in abundance at Skeoch, when we first came. Our predecessor, Mrs Moody, used to put out various tin trays of food and milk for them but they never strayed near the house, preferring the outer reaches of the Dutch Barn. They were a motley collection, of virtually every shade and hue you could imagine. All wild-eyed, very timid and quite mangy. Any overtures on my part, to get better acquainted, were all met with blatant hostility, with much spitting and snarling. I was told, in no uncertain cat terms, that any form of fraternisation, apart from a steady supply of food, was not welcome. They came and went in seemingly endless droves during the first cold months of the year but finally, maybe because of Zebedee claiming the territory as his own, or because of the ferocity of the winter months, they disbanded bit by bit and Zebs was the King of the jungle. His size, it's true, must have appeared quite daunting for the lesser cats. For although a goodly portion of him was his thick black Persian coat, (from the rear he always looked like he was wearing Cossack britches!) he was also extremely meaty. His eyes were the colour of elderflower wine and being a most friendly cat, I was regularly treated to strong displays of affection, usually when I had shorts on. Two large hairy paws giving me a playful 'hug' at about knee level from behind and usually quite unexpectedly. Skeoch for him was paradise, with no traffic and unlimited fields to roam in. He began to depend on us less and less for food, which he preferred to select himself from the wild, returning to us more for affection. He would disappear for weeks on end during the long summer months and just magically appear again, looking sleek, fat and very pleased to see us. I sometimes wondered whether he had a secret 'other life' where he was equally loved and fed by another family but apart from Paddy and Millie (not forgetting King) which didn't seem a likely prospect, there was only Jim and Doreen and neither of them ever fed him. Although we had had him neutered when he was a kitten in London, he had a very paternal disposition to smaller cats and kittens. In fact, during our life in London in the flat, he had been known, on more than one occasion, to bring home various starving strays, through the downstairs open window, leaving them terrified at the bottom of the stairs, while he strolled upstairs and methodically threw sprats, from his supper dish, down the stairs to the waiting stray at the bottom. The first time I witnessed this., I could hardly believe my eyes. There were times when he seemed definitely more human than cat.

It was during our second year that another neighbouring farmer, some 70 or so acres away, arrived round one morning, unannounced, with a small, pugnacious, smooth haired black kitten. He was quit adorable in an 'Evel Knievel' sort of way and we promptly called him Jasper, which suited him admirably. Jasper lived indoors for about 6 months, then, as the weather got warmer, he took to roaming. He had the air of a vagabond about him. Zebedee was ever eager to please him and together they went on lengthy jaunts. Patiently Zebs would allow himself to be pounced on and his very fluffy tail leapt on and 'killed'. Jasper would flatten his tiny body close to the ground, wiggle his little bum and leap on poor unsuspecting Zeb, who was maybe trying to doze in the

sun. There would ensue a mock fight which Jasper was always allowed to win. After one of these long jaunts together, Jasper failed to appear home with Zebedee and no amount of calling or looking for him in the weeks that followed, brought him back. Finally we had to face the fact that Jasper had either permanently left home, or sadly and more probably, that he was dead.

Some six months later, when I was going to the store shed outside the back yard which housed our Cash and Carry tinned goods, I spotted a small, grey, furry object hurl itself from under the shed door and disappear into the undergrowth and nettles nearby. At first I thought it may have been a rat and prodded among the nettles with the handle of the yard broom. Certainly no rat appeared but a very indignant scrap of ash grey fur with huge green eyes. This bit of fluff spat at me and was determined to put up a very good fight with the broom handle. He was one of the prettiest kittens I have ever seen and judging by his size, could have only just got his eyes open. He scarpered away quickly into the undergrowth and I could not retrieve him. Where he had come from I had no idea, as the wild cats that had dominated the farm when I first arrived had long since disappeared.

I put a saucer of milk near the flattened nettles and went back indoors, hoping he would be enticed back. By dusk the milk was still untouched and a further search of the undergrowth produced nothing, but as I turned to go, something made me look at the shed door. It was a very poor fit in it's rickety frame with about a three inch gap at the bottom. Sure enough, peering at me from beneath the gap, I saw the two large green eyes. I knew there were some cardboard cash and carry boxes in there, some of which were filled with old dolls' clothes from the children's toys. I turned and went indoors, knowing that at least for that night our new adoption had a warmish bed. Lancelot, as he became known, never truly took to humans. He would very occasionally allow himself to be caressed but after a minute or two, a quick bite and an unplayful smack with a sharpened paw, reminded you that he was doing you an enormous favour by allowing himself to be stroked and he didn't want any more THANK YOU!

He was quite happy to have the half board and lodging that we provided for him but in the way of all the others before him, he finally took to the call of the wild as well. Sometimes we would catch a glimpse of him in the early evening setting sun, leaping across a field somewhere, like a little grey shadow. He was free and obviously happy and after all, who were we to try and contain him!

CHAPTER 6

As we were more or less forced to take the 'farming lark' seriously, gradually over the months that followed, we bought in batches of two-week-old Hereford heifer calves. A new consignment of calves always lifted our spirits. The majority of them took to bucket feeding very quickly but there were always one or two exceptions and if, after trial and error, one or two refused to drink from the pail, the only recourse was to bottle feed. I never minded this too much, although it was time-consuming. It was lovely to see the look of absolute bliss on their faces when they finally accepted the huge black rubber calf teat, on the end of a large lemonade bottle. Greedily they sucked up the three pints of warm milk, making a froth of bubbles round their mouths and ending up butting any part of your anatomy they could find in their eagerness to find more, when it was finished. Their stubby little tails jiggled about excitedly. Feeding time was great excitement for them and us and their bellows when we arrived in the pens in the morning were deafening. After the milk, they were given clean bedding and the top half of the byre doors were opened to let the fresh air in. They were each given 3 - 4 pints of milk, twice a day for about two months and then gradually introduced to solid food, calf pellets, which was oddly called 'cake' by the farmers, while their milk ration was gradually decreased. They were usually on milk altogether for about ten weeks and by the end of that time they were eating pellets and hay. They would stay in their pens until they were about five to six months old and then would be let out to grass.

It was wonderful to see the display of antics when they were first turned out into a lush meadow. A six month old calf that has been kept inside, is usually quite covered in various layers of deep litter and dung, all of which served to keep them warm during the winter months. With their new found "out of the pen" literally, they galloped around kicking up their heels and discarding manure in all directions, looking rather like gawky school girls, out on a charabanc trip. Then they would just stand and gaze in wonder at the pale spring sunlight and green pasture. It wasn't long before they realised what grass was for and an hour later would see them all munching contentedly away. We always breathed a sigh of relief when we reached this stage, that we had got them through the most difficult stages of rearing.

We now had around 25 heifers, which were to be our mothers. From these we would breed wonderful prize Hereford calves - eventually! They all had names. One we called Fenella (after a certain actress) because she had the most divine long eyelashes I had ever seen on a calf, which she batted in my direction every time the milk bucket appeared. Pansy was so named because she had two sooty patches, one over each eye, making her look rather like a panda bear and so on. One, which we named Flighty, tells her own tale ...

We had been unloading our latest batch of calves from the cattle truck, normally a fairly routine procedure. The ramp was pulled down from the back of the truck and gates placed on either side of the ramp ensured that the calves trotted down, straight into the byres. They normally travelled from Carmarthen, which was a heck of a trek for two-week-old calves but this batch all appeared quite frisky, in spite of the long journey. Finally we got them all safely housed in their new pens, with the exception of the last one, who skidded down the ramp, looked left and right and promptly bunked over the side and away before anyone could grab her. Swift as the wind, she scaled the first wall in sight, which happened to be a jump down to the large bottom fields - and was gone. At this stage we thought we would be able to catch her fairly easily but she was very small and the grass was very tall and above all else, she had about 100 acres she could roam in before she could be stopped.

We started off in hot pursuit, Conrad taking one direction and I the other. We would catch a glimpse of her but as fast as we arrived at where she had been, she had disappeared again. I cannot begin to tell you the hours we spent searching for her. We gave up at about 11.00 a.m., having spent two hours fruitlessly searching. A friend from the village thankfully came up to look after the children. Con and I headed back to the house for a cup of coffee, then back to comb the surrounding area. We snatched a brief lunch and went out again. All this time she was quite enjoying herself. When tired, she just flopped down in the long grass, completely concealed from us until she had rested long enough, or thought we were getting a bit too close.

Hot, filthy and tired, I distinctly heard Conrad emitting a strange sound. At first I thought he was hurt. I saw him, lying on his stomach in the field of sinks. He was in fact 'moo-ing'. About ten yards away from him, the doe-eyed Flighty stood watching him with slight amazement.

"Moo" he called and inched a bit nearer. "Moo, moo". She stood stock still, staring at him in fascination. So did I, doubled up in silent laughter. I didn't dare make a move in case I startled her. Conrad crawled on, encouraged by what he hoped were motherly cow sounds. He inched nearer. Flighty stood her ground. "Mooooo," he called very softly. He was almost upon her, gaining ground. Suddenly, head down, she felt he was near enough and off she took, sailing over yet another dyke wall and into the woods at the bottom.

The air was more mauve than blue when Con got to his feet and we raced off again. I didn't even have time to congratulate him on his performance! The sweat by now was pouring off us; our limbs were aching and the flies, bracken and brambles inhibited every movement. We had been tracking her off and on for nearly seven hours. She had not had anything to drink and we hoped that she might be getting exhausted herself - but no such luck. We were most worried that night would fall, without having found her. Lost and alone, she could wander anywhere.

Eventually we sighted her some distance on. She had managed to get herself under the barbed wire fence at the bottom of our wood, which marked the end of our boundary and commenced our neighbour's farmland. We didn't know this terrain at all well and there were areas of it that were exceedingly marshy with large patches of spiteful blackthorn. It went on for miles. Time was ticking on and she was getting further into the wilds. Catching up with Conrad, we stopped for a moment to catch our breath.

"I have an idea which just may work," he said as we flopped down, exhausted.

"Let's get our older cows down to the bottom fields. If she sees or hears them, she may make a beeline for them, thinking perhaps her mother is with them".

What a brainwave! We sprinted back to the farm with renewed energy, opened up the top field gates and shooed the cows down to the bottom pasture and waited. Sure enough, after 15 minutes or so and with much mooing, out she trotted, bellowing indignantly at our herd. Like so many Aunties, they rushed up to her, nuzzling and licking her. This was great. She was perfectly content. Conrad reckoned that if we drove the cows into the bottom byre, she might go in with them and we could halter her, once safely inside. Gradually and very unhurriedly, we started driving the herd towards the bottom cow shed. To the side of this shed was the large, slushy green manure heap which I had encountered when we first arrived and got stuck in when pregnant, chasing Zebedee. The cows used to wade through this quite unconcerned and the first few started walking through, into the byre. Flighty trotted along quite happily with them, until she reached the midden. Con and I were right behind. Suddenly she realised the trap. The cows had disappeared inside and she glanced nervously around, ready to take flight again. She turned, skidded and tried to make a bolt for it, at which time I launched myself into the air and made a desperate rugger tackle on her, wellington boots and dungarees in all directions. We both landed in the manure heap - (again for some of us!) with her kicking and bellowing furiously in indignation and surprise more than hurt. I hung on for grim death until Conrad grabbed her and finally haltered her. We both emerged smelling atrociously but I was triumphant. Flighty was taken to a "high security pen"! We thought a few days in solitary confinement might calm her down. She was washed down, fed and given a bed of extremely deep straw. Certainly she didn't seem a jot worse for her ordeal. It was only Con and I, relaxing in a hot Badedas bath, nursing a large whisky and watching our bruises turn a lovely shade of purple, who were the worse for wear but we did at least have the thought that we had got the flighty little blighter in! In fairness to her, all young cows are nosey, adventurous and completely uncontrollable. They are curious to know 'what is through this gate?' 'What is over the other side of this wall?' 'Does this fence give when I push against it?'

On one occasion 'the other side of the wall' happened to be our garden! We arrived home from shopping in Dumfries one morning to a lot of mooing and uproar at the rear of the house. Charging round the back, our worst fears were confirmed. Leaping around the garden, in a state of total euphoria, were eight or ten nine-month-old heifers. They had eaten most of the broccoli, snatched mouthfuls off the Syringa and left their nice, large brown cow pats all over the terrace steps and lawn. Most of the ground was poached up and what hadn't been eaten, had been trampled. How none of them had ended up in the pond, was a miracle. We had to remain calm and hopefully, gently, get them back through the wall they had happily knocked down. It was impossible to guide them through the garden gate, as it was far too narrow and they were already in a highly charged state. Like a party of naughty school girls who had been discovered playing truant, they eyed us warily, snorting and head butting. Having demolished a fair portion of the vegetable garden and with some of their exuberance now starting to wane slightly, we raced back to the barn and poured lavish amounts of cattle feed into a couple of pails and returned to the garden. Greedy noses twitched, eyes lit up. During their six months in the byres, pails were synonymous with food. They crowded round, smelling the pungent molasses aroma. The trick now was to somehow get down the garden path with the buckets, through the broken stone wall and hope they would follow. Like a juvenile hockey team, they jostled each other for the first place, the contents of the buckets their one goal. With a bucket each, we sped down the garden path in a somewhat hazardous fashion, like the Pied Piper of Hamlin, and we finally succeeded in getting them back to the bottom field, distributing the contents of the buckets on the ground. Wearily, we went back up to the garden and surveyed the chaos. Mending the dry stone wall was obviously the first priority to avoid the same thing happening again. The garden was a total mess. All the nurturing of seedlings, the patient weeks of planting and cultivating - now looking like the Somme! Oh the Good Life, the freedom.....the work!

CHAPTER 7

The months ticked by and when the herd was about eighteen months old, not having a bull of our own, we had them all artificially inseminated with prize Hereford bull stock, from the man who was, for the want of a better word, the 'AI man'. The cows grew fatter and sleeker and in the prime of motherhood, especially Lu-Lu, our first born, who seemed to blossom as her stomach got bigger. We were so pleased. It started to look like the dream was becoming reality. That after all the problems that beset us during the last eighteen months, things were turning out worthwhile after all. We had nursed and watched our little herd, seen them through various illnesses and above all, learned a great deal. A far cry from the early days of feeding Lu-Lu with porridge! We were beginning to think and feel like farmers and had earned a modicum of respect from our neighbours. It was a tough part of the world and a tough business to take on when you know nothing at all about it.

Brucellosis (contagious abortion) had been rife in some parts of the neighbouring farms and obviously there was a lot of concern about eradicating it from the area. It had become compulsory for farmers to buy only accredited herds from areas where Brucellosis had been wiped out. Calves were inoculated against it and every precaution was taken. Our own pastures had been 'rested' for some considerable time before we put any calves out and we had been very careful to follow all the precautions involved in the Brucellosis eradication scheme.

It came as a shock therefore, to discover one of our fit and healthy, pregnant cows, had aborted. We went up to the top field where she was and looked at the pathetic foetus of a four month old calf, perfect in every detail. It would have been a fine calf. With heavy hearts we went back indoors to ring the man from the Ministry of Agriculture in Dumfries. He arrived in a cloud of dust, green wellies and a Vauxhall Viva. He took samples, tests, tubes of this, phials of that and told us he would be back in touch within the next few days. We were worried. our pasture was considered 'clean'. We had bought only accredited calves and done everything by the book.

The interim period, waiting for the results of the tests, was torture for us. It occupied our minds every waking moment and through most nights as well. Anxiously we peered at the cows for signs of illness. We watched them constantly and reported to each other, almost on an hourly basis. We prayed that the earlier abortion was just a one-off situation. The waiting seemed interminable.

The ministry man finally returned some days later. He was grim. We could tell by the expression on his face that the news was not good.

"That's it, I'm afraid," he said. "The cow that aborted has Brucellosis".

We were dumb struck.

"What exactly does that mean?" we asked.

"I shall have to test every one of your cows, to see if they are reactors," he said. "If they are, they will have to be put down straight away".

"How did she get it?" we asked. "We've done everything we can to prevent it".

Sadly he shook his head.

"I know," he replied "but you see, there is so much of it around here. It would only take a fox to run across a pasture which was infected and then on to your land, or a crow that has picked up something from a neighbouring farm that was infected and deposit it on your land and you've got it. I am so sorry."

He arranged to come back a few days later and give the herd the first of three tests. With total despair washing over us, we arranged to bring them into the byres.

The cows meanwhile, blissfully unaware of their fate, got fatter. They had the three tests. More waiting. We were told to keep them away from any other calves or young heifers. More waiting. More sleepless nights, more anxiety. The work, the hope. It was a nightmare. It dominated our minds constantly. Surely there was some mistake. Finally the day dawned when we had the results. With the exception of about four cows, the remainder had to be slaughtered. Not because they were all reactors but because even though they had been negative to all three tests, they had been in close contact with reactors and there could be no risks taken. My lovely Lu-Lu was one of those that was not a reactor but still had to be put down.

We were told we would have compensation at the

current market price for them ....it was poor repayment.

They were grim days when batch by batch, we loaded our fine herd into the lorries to take them to the slaughterhouse. Through a veil of tears I watched, as the trucks pulled out of the yard at the top of the lane. We heard their bewildered mooing all the way down to the bottom!

We were desolate. What a cruel twist of fate to have given us so much and then snatch it back again. That was the end of the breeding herd. What to do now?

CHAPTER 8

The terrible outcome of Brucellosis left us in a state of total shock as you can imagine. It made us yet again take a serious look at what we could do to make a going concern of Skeoch. With all our hard work of the last two years completely ruined, it took us some time to decide what we should now go for. We spent hours discussing the pros and cons of other ventures. Just turning the fields into arable land for hay or barley was not a viable proposition. If the summer was bad, the crops would be ruined and our livelihood with it, overnight. We had to do something that would not depend entirely on the whims and unpredictability of nature, to survive. Just letting the land for grazing was no answer either, as there was so much land up there and the financial return was poor.

We spoke at length with other farmers who had witnessed the same problems among their community. They were kind and sympathetic, offering various advice. One thing we learned, was that Brucellosis affected only heifers and cows, so young bullocks were not at risk.

After much heart searching we scraped together the last of our dwindling resources and with the compensation

money from the Ministry of Agriculture, we decided to try again. This time buying in two week old bull calves from accredited herds and rearing them to 18 months, hopefully to sell on as 'store cattle'. At eighteen months they would be bought at market by other farmers, who would keep them at grass for a further period of time, until they were a good size. But all this was further expenditure, with nothing coming back into the kitty in the meantime.

The calves we were to buy now would not be the brown and white faced Herefords but the little black Aberdeen Angus. Sturdy and robust and reportedly the best beef in the world. They were stocky and very pushy little calves with healthy appetites.

Conrad was now feeling much fitter and healthier. The black eye patch that he had been forced to wear to retain his balance, whilst making him look like a rather dashing pirate, thankfully ditched. The double vision and balance were completely restored to normal and the illness itself, left as mysteriously as it had come. For that we were delighted but having been told that he would never again act, had been a major blow. Psychologically as well as financially. He now felt, more or less, completely normal, except when he was over tired, then his eyes would start to play up and he would experience vertigo. This eventually passed if he rested up and took things a bit easier.

One morning his Agent in London rang. She asked how he was feeling. Did he think he could tackle acting again? Would he like to have a go? His name had been chosen to play the leading role of Rochester in Jane Eyre in Worthing for the theatre. It was The Connaught Theatre in Worthing where he had originally spent a season as a fledgling actor. The thing was - could he do it? It would be the first time since his recovery that he had acted and apart from sustaining the leading role, it would also be in front of a live audience every day. There would be three weeks rehearsal and four weeks playing, with evening performances and matinees. It was a tough assignment and an exacting script but he had to prove to himself that he could still do it. It could be a GREAT breakthrough and give him back the self confidence that is so necessary to be a stage actor. He decided to accept - with just a bit of apprehension! But the dye was now cast and the script arrived within the week. He worked incredibly hard on the production and when the play finally opened, the children and I took some time off and went down to Worthing for a few days. I was so nervous for him. I knew that he, being him, would brush it off. No problem, he was fine. I also knew that with the bleak future forecast given to us, regarding his acting capability, so much hung on whether the play was a success or failure. A lot depended on him and he had an awesome responsibility, not just for us but for the rest of the cast as well.

Con had arranged for us to have a box at the theatre and to come to a matinee performance, as the children were still very young. The play went very well - then came a dramatic pause in the dialogue, a crucial moment between Jane Eyre and Rochester, when she discovers he is blind and he has discovered she has come back - you could hear a pin drop.

"That's my Daddy, that is," Sarah's small voice rang through the auditorium. Talk about stop the show! It was a wonderful moment and one I think I can safely say Conrad and I will never forget. How he managed to continue, I don't know - but he did, to a thunderous applause. A write-up in a local Worthing paper mentioned him as "the definitive Rochester". He was back!

About a week before we knew of the offer of Rochester, we had bought in another batch of two week old calves and among them was a little brown and white Simmenthal calf. They all appeared healthy and all had extremely good appetites, so there was no undue concern that Conrad would be leaving me with anything I couldn't handle, during his first rehearsal period. By now I was quite an old hand at mixing vast quantities of the powdered milk and warm water needed for calf feed and the whole procedure of feeding and bedding the new arrivals posed no problems.

It was with some alarm then, that one morning a few days after Con had left, that I went in to feed the Simmenthal, only to find him listlessly staring into the corner of his pen. He refused all attempts to feed him; his head hung down and he was breathing rapidly. I tried several times during the morning to get him to drink but he just stood rooted to the spot, staring into space, oblivious to anything I tried. By early afternoon I was extremely worried about him and called our veterinary surgery.

The practice - Mr Allison and Mr Murchie - was a very interesting partnership. Interesting because the two vets were totally unalike in their individual approaches to sick animals. Mr Allison was a real, solid Scots farm vet. Extremely good, extremely practical. No frills, no fuss. Mr Murchie on the other hand, had an "animal bedside manner". He was altogether softer in talk and action. Together they made a wonderful team. You needed Allison for your cows and horses and Murchie for your dog, cat or bird. That is not to say he was not a good vet, on the contrary, but his nature seemed more in line with the less brutalising aspects of being a vet. Actually, my very first meeting with Mr Allison was, on reflection, a rather embarrassing one. Embarrassing for me, that is, and it came about like this ...

During August, in our first summer at Skeoch, on one Thursday morning, the sun was already up hot and strong. I felt in a lazy, slightly indulgent mood and wanted to spend some time sun bathing and recovering after the birth of Sarah. We had no livestock then to worry about and so I dragged an old sun lounger round to the back of the house, overlooking the rear garden and terrace. It was a fantastic sun trap and of course, with no neighbours - totally private. I stripped off, down to my briefs and lay sun worshipping, topless. The postman Andy had already delivered the letters, so we were not expecting any disturbances. The sun blazed down, the birds sang and all was tranquil, until a slightly nervous cough at my side revealed a rather red-faced Mr Allison.

"I'm so sorry," he muttered, extending his hand. "I'm Mr Allison, the vet in Dumfries and I've just been to see Jim Kissock. I heard you were new to the neighbourhood and came up to pay my - er - respects". I shot off the sun lounger, covering my boobs with one hand and shaking his proffered hand with the other. We then had a rather ridiculous conversation for about five minutes, trying to get over our mutual embarrassment, while pretending nothing unusual was happening. I stood there, very conscious of my rather tired, off-white briefs. You know the ones, those that get washed accidentally along with the blue wash! Finally we finished our pleasantries, just as Conrad came outside bearing a tray which had two large gin and tonics, which he had thoughtfully prepared as a pre-lunch aperitif for us. Mr Allison shook hands with Conrad, got in his car and whizzed back down the lane. Neither of us had heard him arrive, poor man - but that was our first meeting.

Back to the sick Simmenthal. I got Mr Allison on the phone this particular afternoon and as ever, he wasted no time in arriving. Striding out to the calf pen, exuding an air of being totally in control, I immediately felt calmer and reassured. After a few minutes he strode back to the house, carrying his large black bag. He opened it up and took out a load of hypodermics and antibiotics on the table. "Right," he said, "I've had a look at him. He has meningitis". I was aghast. "Does he have to be put down?" I asked. "Maybe not, if you do as I tell you," he replied. "Keep this formula in the fridge and give him 5 cc's into the muscle every four hours for the next four days. I'll be up to see him after then. Ring me if you have any problems or if he gets any worse". With that he smiled and started to make for the back door.

"Just a minute," I mumbled. "I've never given an injection in my life before. What muscle, where?" He grinned. "Ach, just any muscle will do. Just stick the needle in hard and push the plunger. It's really quite simple". His blue eyes twinkled. With that he strode back to the car, bid me a hasty goodbye and left. I returned to the kitchen, feeling most apprehensive.

By the time I had to administer the first injection, I couldn't stop my hand from shaking. Uneasily I got the formula from the fridge and loaded a syringe, squirting a bit out of the top as I had seen it done countless times on "Emergency Ward 10", an old TV series. So far so good. I wished with all my heart that Doreen was around. She would have done it with no problem at all but she was at work in Dumfries. God, I was such a wimp! I walked to the calf pen and looked at the frail little body, still standing in the same place, head still hanging down limply. He didn't move or even seem aware of my presence. With shaking hands I felt him all over, trying desperately to find out what was muscle and what was bone. I was terrified of sticking the wretched needle in the wrong place. Finally I located what I thought was a suitable place and with a silent prayer, stabbed the needle in and pushed the plunger. I was amazed at just how tough the calf's skin was and equally amazed that he didn't react in any way. Flushed with success, I covered him with hugs and rubbed along his bony spine. I had done it! Next time it would be easier. His life depended on it. By the end of the fourth day, I was positively blasé and constantly, boringly told Conrad about it on the phone. The calf responded and thank goodness, by the time Mr Allison arrived back up, we were over the worst. His appetite returned with a vengeance and he made up for lost time. The little Simmenthal eventually grew into a very large Simmenthal and fetched an extremely healthy sum at market, some eighteen months later.

In fact market days were a cross between excitement, anticipation and dread. I had never been to a cattle market before until we came to Scotland and in the early days, before we had anything ourselves to sell, we would occasionally go with Jim Kissock. The action there was amazing. Noise, bustle. Pens of cattle, sheep, pigs all bellowing at the same time. Some animals looking so pitiful that the knackers yard was perhaps the only place for them. But in the ring itself, the Auctioneer, smacking down his gavel every few minutes and the bidding going on so fast, that I had great difficulty following at all. Take into consideration that the Auctioneer spoke with a very strong Scottish brogue on top. I was fascinated. The old hands that stood round the sides of the ring, would twitch an eyelid, pull an ear, tap a nose. You never actually saw anyone bidding. It was like being on some Secret Service initiation rite but the Auctioneer's gimlet eyes never missed a trick. I held my breath and didn't even dare look up for the first half hour, in case I had inadvertently bought six Highland Longhorns! The old sage farmers, dressed in patched tweed jackets that had seen much better days, stood close to the ring. In fact we later discovered, you never went to an Auction in anything but your tattiest clothes. When Con and I arrived on the first occasion, the whole place seemed to go quiet. Eyebrows were raised at my new green wellies and jeans and Conrad, I thought looking every inch the gentleman farmer in his new anorak, was met with some slightly bemused stares. Obviously this was NOT the gear to wear to the Dumfries cattle market.

The great gate clanged back, a load of shouting and much thumping of sticks - bellows and snorts as various collections of cattle of various shapes and sizes, ages and breeds, would come pounding into the ring, to be driven round by some rather brutish lad, with a stout stave in his hands. The poor creatures galloped around, completely disorientated, some of them having travelled for quite a few hours in the cattle trucks before they got there.

Depending on the weight of the animals, the price per hundredweight at the market on that particular day, one ended up despondent or not. At the end of the bidding we always went to the nearest pub, either to celebrate with Jim or console him. One way or another, quite a proportion of the sale proceeds was spent in the Castle Arms afterwards. Sometimes we even felt that we were starting to belong to the farming community. It was such a far cry from the life we had known before; for Conrad it was a very strange time. The twin lives of acting and farming were so opposite. It was all very well to act at being a farmer, like "Bill exits stage left to the cattle byres" but BEING a farmer AND an actor, is another thing. Neither fish, fowl or.... ? What a strange life. For me it was much more straightforward. I had the best of both worlds inasmuch as I was a wife and mother to two little ones; a surrogate mother to the calves and I suppose you could call me a 'farmer-ess'! I wasn't the one, thank goodness, who had to have a split focus and tear back and forth to England.

In the years that followed, Conrad of course, became a

dab hand at giving cattle injections. Every year the cattle had to be corralled and given Nilverm injections, which protected them from worms from the pasture. This was no mean feat. The cattle had to be rounded up from the various fields and brought into the yard. One by one, we had to get them into a very ancient "crush". This was a narrow wooden construction. Barred across each side and with front doors, with just enough room for the cow to put its head through. Once it was driven in place, we put a huge metal pole across the front doors, to prevent them opening. The cow was now effectively trapped, being unable to turn round or back out. When it was firmly in position, Conrad would shimmy up the side bars, bend over the, by now, very irate bullock and administer the injection. Much raging and bellowing (both from the cow and ourselves), much swearing and many bruises, usually followed. I stood at the front end, ready to move the iron bar and open the doors to let the cow out. I had a few side kicks on the knee during the process! We dreaded doing it every year but it was a necessary precaution and we were very thankful when it was over. It wasn't, after all, for wimps. It was a hard, unyielding life, - 24/7 life really.

Looking back, I can't believe how timid I was when we first started farming and how quickly that evaporates, when it becomes your livelihood. Conrad and I and the children would stroll through a herd of large bullocks in the fields, move them, walk among them, inspect them, with no fear. When they were inside, from the age of two weeks, until they were out to pasture at five months old, they were all totally individual to us. We recognised them all and knew all their individual foibles. After two months on the grass, they were unrecognisable - we could no longer tell one from the other. It was a necessary distancing.

CHAPTER 9

The entry of 'Oscar' into our lives, was another one of those delightful incidents, which seemed to happen periodically. Perhaps it was the reputation we were fast building, for taking on lame ducks!

My friend Doreen, had taken me out lambing one freezing day, with her lovely black and white border collie, Pat. She was a remarkable dog and knew instinctively what was expected of her. She was extremely intelligent and kindly and it was fascinating to watch her in action. If Doreen spotted a ewe that looked as if it was going into labour, she would give a low whistle to Pat. Quietly, without distressing the sheep, Pat would encircle it, weaving back and forth until the ewe was successfully separated from the rest of the flock, usually penning it carefully in a corner of the field, until Doreen came. If the sheep started to get nervous or show signs of aggression, Pat would take a sprint and hold the sheep by taking it's top lip in her mouth, exerting just enough pressure to stop the sheep from trying to escape - but this only happened on very rare occasions. Swifty and deftly, Doreen would gently pull the creature to the ground. With years of practice behind her, she would gently ease her small hand into the sheep, carefully manipulating round the tiny hooves and head of the emerging lamb. Sometimes, when there were twins or triplets, the procedure was quite complicated, trying to sort out whose limbs belonged to whom. But all seemed to be achieved with incredible calmness and efficiency. Doreen confessed to me that it was one of her favourite jobs, in spite of the freezing temperatures and totally unsociable hours, cramming it all, somehow, into her daily life and holding down a demanding full time job in Dumfries as well. Her patience and energy were amazing.

It was March and we had had a filthy night of gales and sleet, which had battered away into the early hours of the morning. By about 4.00a.m. the wind suddenly abated, leaving us calm but cold. We woke up to a hoary white, frost-laden day, with piercing blue skies and diamond brightness. Conrad and I had just finished the morning battle with the calves, trying to train them to drink without biting our fingers off and for those that flatly refused, bottle feeding. Either way, we usually ended up with more milk down us, than them!

The telephone jangled shrilly on the outside bell and we dashed indoors to answer it. Jim's voice came thickly over the line.

"Do ya want a 'silly' lamb, Jennie? It's one of triplets an' more dead than alive. If ya care to come doon, ya can see for yersel".

We always enjoyed the challenge of Jim's rather fatalistic 'more dead than alive' statements, which preceded nearly every piece of livestock we took from him. Retrospectively, though, he was probably nearly always right! We wasted no time in jumping in the van and steering down the lane to his farm.

The front room of Scour Farm always had a fire lit, summer or winter, and this morning was no exception. It blazed merrily away in the small black grate and in front of it, lying on some soggy newspaper, getting more and more scorched, was the tiniest lamb I have ever seen. It was all legs and head. It's eyes were tightly closed and it was shaking with ague. (Sound reminiscent?) The back of it was still absolutely sodden. It lay there, steaming in the front, looking pretty dead really. A more pathetic creature you couldn't imagine and if I can say that half a pound of bacon looked positively robust by comparison (in fact it weighed no more than 10 ounces), that will give you some idea of its size.

I bent down and picked it up. The head lolled to one side, the eyes stayed closed, it was barely breathing. Jim told us that it had been born very early in the morning, in some of the worst weather, the sleet, which had finally turned to hard frost. Although the mother had accepted her other two offspring, this one was the 'silly' lamb. In other words, the runt and she had flatly rejected it. Plainly, unless it had some nutrition, it was definitely not going to survive. In fact I doubted it would anyway but it was worth a try. Jim's fatalistic attitude was becoming contagious! As usual, Jim came to the rescue with a small bag of powdered lamb's milk which he handed over with a grin and "See what ya can do wi' that Jennie!"

Happily I wrapped the lamb in a discarded old woollen jacket of Kate's, thoughtfully grabbed at the last minute before we left home, and we skidded our way back up to Skeoch.

Once back inside, I unwrapped him. There seemed very little sign of life in him and the first thing we did, was to rub him, quite firmly with an old towel to dry out his frozen woolly back, trying to massage some warmth and circulation into him. There was a slight response but not much. We enlarged the hole in one of Sarah's bottle teats (again!) and put a small amount of glucose and water into a tiny bottle. I prized the frozen lips apart and we forced the teat into his mouth, squeezing his jaws together with my other hand. He coughed, spat, swallowed and finally, after about 10 minutes of struggling, we succeeded in getting him to suck. With eyes still closed, he drank a small quantity of the mixture. Carefully we laid him in a cardboard box, lined with more woollen discards and a hot water bottle and placed him in Sarah's playpen, in front of the central heating radiator.

About half an hour went by before the eyes finally opened, the head drunkenly lifted and our first lamb took a first look at his surroundings. He gave us a bleary look and a half-hearted bleat and tried, unsuccessfully, to get to his feet. We watched as first the front legs came up but as soon as he tried to get the back ones up as well, the front ones promptly buckled under him, like a pantomime horse. This see-saw process took him about 15 minutes until he swayingly mastered the art of getting all four legs co-ordinated together. We watched fascinated and elated. So far so good - but walking was a different matter altogether!

Oscar, as we had decided to call him (I had just finished reading the complete works of Oscar Wilde) gained strength at an amazing rate. He demanded to be fed every 3 - 4 hours, bleating pathetically if he felt nourishment was overdue. Because I fed him, he automatically presumed that I was 'mother'....Jennie had a little lamb....how true. He soon got the hang of squeezing between the bars of the playpen and was quite content to just follow me around the kitchen or wherever I was going but as soon as I was out of sight, he set up the most robust bleating ever. If I went to the loo, I had to keep the door open, so that he could still see me and on the first occasion that I decided to do the decent thing and lock the bathroom door, he decided to battle charge it, to force his way in. The tiny hooves slipped on the cushion floor in the kitchen and he persistently took a flying leap at the loo door until, fearing he would damage himself, I reluctantly had to let him in. For one so small, he was incredibly demanding - you could say he was a third child in fact!

His bottle feed lasted what seemed an interminable three months until we decided that really he was a lamb after all (and could not be house-trained...) and Conrad said he would be better off in one of the warm outhouses, surrounded by straw bales, until I went out to feed him. Carefully we put him in the outhouse, despite enormous protest from him and we put a good eight bales of straw around him so that he was effectively penned snugly in, and locked the door. He was warm, dry and cosy, with plenty of light and fresh air. But Oscar had different ideas on this altogether. His bleating went on relentlessly, hour after hour. We resolved to stand firm and tried not to worry. After about 3 hours, just as it was getting near feeding time, we noticed that the bleating was getting less and less frequent and we hoped that he was getting used to his new establishment and this new discipline. Not so. When I finally went across the yard with the bottle and opened the shed door, what a sight met my eyes. He had determinedly butted his way through the straw bales, knocking them all over. Having achieved that, he then appeared to have battle charged the shed door, using his nose as the ram. Poor Oscar. His nose was now three times it's normal size, having swollen up to a red, raw blob. He bleated at me pathetically. This would never do. I fed him and brought him back into the house while we discussed what to do. Clearly, keeping him in the shed was not an option. He had to be somewhere where he couldn't damage himself. In the end we decided that the only thing we could do was to put an old dog collar on him and tether him to the end of a longish rope, tied to stake on the lawn, in front of the kitchen window. It was now high summer and warm - and I could keep an eye on him from there. Gradually he adapted to his new surroundings, although to start with he gave a half-hearted bleat every time I was in earshot. But he started eating grass and I was able to cut down his feeds until he was finally weaned. Since he was a little ram, Jim suggested that he had better come up one afternoon and castrate him. I wasn't looking forward to it much but a few weeks later Jim appeared, armed with... 2 thick elastic bands. One he wound tightly around Oscar's tail - the other went round his 'bits and pieces'. I drew the short straw of having to hold him while this was being done! The whole process would take about 8 weeks. So, when you see spring lambs leaping into the air, it isn't necessarily for the joy of living! I must say, he didn't show any obvious signs of discomfort and some weeks later, on a warm sunny afternoon, Katie went out into the garden to play. Today, she decided to play with Oscar! Suddenly there was mayhem. She let out a yell and ran back indoors howling, tears streaming down her little flushed face and the end of Oscar's tail held firmly in her small hand.

"Mummy," she wailed, "I was only playing with him and his tail came off in my hand. I've pulled his taioool off". She looked at me amazed as I started to laugh, and wiping her tear-stained face, I told her that he was perfectly all right. She took some convincing and it was not until I persuaded her to come with me into the garden and see him grazing away quite happily, that she finally believed me. He gazed up at us, quite unconcerned, his tail now neatly docked. Fortunately we never found the other parts of his anatomy in the garden, or I really would have had some explaining to do!

When Oscar was a year old, we thought that, much as

we were fond of him, he would have to go to market. A rapidly growing sheep, tethered in various parts of the garden, was not really productive in the farming sense of the word, except as a portable lawn mower and we could not, obviously, keep him for our own consumption, so we made a plan for him to go to market where some other farmer could take him and put him with a flock. On the last day of the sheep sales in Dumfries, Conrad ruefully got up to do the deed and load Oscar into the van. Oscar was not to be found. Oscar had broken his tethering rope (something he had never done before). Oscar had, in short, disappeared. After fruitless searching in all the likely places, Conrad finally found him hiding in the pig sty. Perhaps he knew it was the last day of the sales and if we missed that, he would have been reprieved for another season. It was with a rather heavy heart that I watched the van drive down the lane with Oscar inside. He fetched a very low price at the market!!

CHAPTER 10

Helen Sherry ran the local Post Office-cum-general store. It was situated about one and a half miles from Skeoch and provided the villagers of Shawhead with all the usual necessities. Helen was a small, dark-haired woman, with huge, luminous brown eyes. Her face, in repose, was quite stern, a 'no-nonsense' sort of face but when she smiled, she radiated kindness and warmth. I think in her youth, she would have been a very beautiful woman. Hidden behind this rather straight exterior lurked, in fact, a wonderful sense of humour but it was kept in reserve until she got to know you. Her age was difficult to assess, not being old, but she was somewhere, I suppose, between 35 and 40. She was also the local Registrar for Births, Deaths and Marriages. We still have the birth certificate for Sarah, written out in her beautiful copperplate handwriting, announcing, somewhat romantically, that Sarah Ann was "Born in the Parish of Irongray".

The Post Office, along with the accommodation attached to it, shone like a new pin. I remember the first time I set foot inside Helen's kitchen. It dazzled, it gleamed. Even the coal in the brilliant copper hod, looked polished. Her ancient cooker, sparkling as new, like some ritzy advert for Homes and Gardens. I have to confess, I beetled back home afterwards and guiltily scrubbed away at everything in sight for all I was worth. As usual, this zealousness was not to last long and I was back to burnt milk on the stove and mud on the carpet within a very short space of time.

The little village of Shawhead had a cluster of houses, the village school and also a village hall. On the 31st October every year, there was a children's fancy dress Halloween party held there. Every child in the radius of about 10 miles was expected to attend, with parents in tow. It was the BIG event of Shawhead each year and costume-making commenced months in advance. Yards of colourful fabric whizzed through sewing machines, papier mach armatures burst out of bedrooms, masks, capes, hats. It was a huge test of imagination and inventiveness and was taken very seriously by all.

Our first year there with Katie, saw me on the sitting room floor at Skeoch, cutting out huge cardboard wings - she was to be a butterfly! This gorgeous multi-coloured creation eventually looked superb when harnessed on to her - the problem was getting them into the car in one piece. She did get first prize though. Prizes in fact, were given for every age group, every category and no child went home without some sort of prize. After the judging and dancing, huge plates of 'champers' were dispensed to children and adults alike. This was vast saucepans of mashed potato, with huge quantities of grated Scottish cheddar cheese melted into it. It provided a formidable layer of insulation for the brave-hearted, like one little chap who arrived dressed only in balloons, calling himself 'Inflation'! Cakes, crisps, apples, toffees, lemonade and coffee were also served out and the evening generally broke up at about 11.00 p.m. when tired, tearful small faces started getting truculent. A splendid Halloween night, which was repeated every year with both Kate and Sarah, who went as a little red devil when she was about two and a half and walked away, again with the first prize.

So the days dwindled down and ..."it was my thirtieth year to heaven". A blisteringly cold January day. With a screech of tyres, the battered old Chrysler of Mrs Moody swung to a halt in the yard. She got out of the car and marched her way down to the cottage door. She was always re-assuringly the same. Good old brown tweed coat, tied with twine. Short hair cut, gripped back at the side and carrying her large, dog-chewed handbag, which contained her 60 ciggies a day. Under the other arm was ,of all things, a miniature apricot poodle, called Guy, which she absolutely adored. She was of ample proportions, with a very kind face and extremely twinkly eyes. She spoke always in a mock authoritarian voice. No frills, no fuss but a lot of tongue-in-cheek humour. She strode into the kitchen, giving me a gruff kiss on the face.

"Heard it was yer birthday, so I got yer a wee something. It's in the car," she said. "I"ll have the coffee first though". We sat and had coffee and generally chatted while she lit up one of her many cigs. She took a keen interest in all we were doing at the farm and she still retained a few acres of Skeoch as her own. Finally I accompanied her to her car. From the boot there issued forth the most peculiar sound. She unlocked it, grinning at me, and swung it open. Inside was a large wriggling sack, tied at the neck with bailer twine. "There it is hen," she smiled. "Happy Birthday. I thought you'd appreciate something useful".

With difficulty we hauled the moving, snorting sack out of the boot. "Better open it in the field by the sties," she grinned, and with that, gave a cheery wave and was gone.

We carted the sack to the field and opened it out. Two tiny, very irate piglets trotted out, making an incredible amount of noise. One was all pink, the other pink and black. Useful? Well, having made the mistake of letting "Pinky and Perky" free range to root, in the small paddock by the side of the house, they spent the whole of the first few winter months carefully digging up the oil pipe to the central heating tank. Like a pair of archaeologists, they meticulously uncovered a section a day, looking extremely pleased with themselves, their faces and snouts covered in earth.

They grew very quickly and grunted continuously to each other. However, the one thing guaranteed to dispel their good nature, was cleaning out the sties. This they loathed and detested with a passion. They were extremely proprietorial about their bedding and whenever I undertook the unenviable task of a 'good clean-out' it was a permanent battle of wills. They stood in the way, lay in the way, squealed as if they were being horsewhipped and generally showed their displeasure noisily, even preparing to do battle with the pitch fork on occasions. When they were finally cleaned out and fresh bedding put in the sties, they would immediately hole themselves up for about 3 days, chewing the straw into chaff and spitting it out again, until their bed was of a satisfactory consistency and depth. Once this arduous task was completed, they were all sweetness and light until the next time. Inevitably they got very large and trying to be an efficient, self-sufficient farm, we hardened our hearts and sent them away to the abatoir. They would keep us in pork for some considerable time and, as Conrad and I constantly reminded each other, you cannot keep two fully grown pigs as pets.

Four halves of pork finally arrived back at Skeoch. Conrad butchered them on the kitchen table and I put it into freezer bags and labelled it up. " Two and a half pounds - Perky - roasting joint". "Casserole - Pinky" etc. It was a good six months before we could eat a piece of pork. Not as bad as my childhood though, when my parents kept a large black pig named Billy, who adored having his stomach tickled. Finally my parents decided that it was time for Billy to go to the Sty in the Sky and he came back quartered and smoked. The first morning my mother served up fried eggs and rashers of smoked Billy, NONE of us could eat it. Eventually Billy was buried under the apple tree in the garden, in three quarters and various rashers!

Along with our slaughtered pigs, 2 large black bin liners were unceremoniously dumped into the van. When I got them home and opened them - there staring up at me, were the heads of the two pigs. It was their final moment of triumph, at my reaction.

"It's for to make brawn with," my Scottish friends told me. I couldn't face this - it was the last straw.

"Give them to me," Conrad masterfully said. "I'll dispose of them". My hero. He disappeared out of the house with the offending bags and re-appeared a few minutes later empty handed. Trying to sound casual, I asked him what he had done with them.

"I dumped them in Fairyland" he said. 'Fairyland' was the wild spot by the side of the house, which in the Spring was covered in snow drops. The children had christened it Fairyland because it looked so pretty. At that moment it was three feet high in nettles. Unfortunately, much later in the summer, when the nettles had died back, it looked more like 'Dead Man's Gulch'. We did the decent thing and cremated them before the children noticed!

Never again did we indulge in pigs. Apart from eggs, it really was very difficult to eat one's own meat produce. Our local farmer, on occasion would rent our two top fields from us, a total of some 18 acres, to graze his sheep on. In return we charged no rent but were assured three sheep for the freezer as and when we wanted them. This form of barter was much more preferable.

Being in the South West of Scotland, we were favoured with the Gulf Stream passing our side, bringing with it occasionally very mild weather but equally, too, we seemed to have more than our fair share of snow. Sliding up the steep slope that led from our back door to the gate and across the yard to the cattle byres, was extremely hazardous, especially when trying to carry two large buckets, full to the brim, with steaming hot milk for the calves. Invariably some sloshed down a wellington boot on the way. Worse than this though, were the gales. If they happened to be married with a snow storm as well, they made a formidable combination. Cattle that wintered out were fed hay twice a day and if it was snowy, cow pellets. It was a constant chore lugging bales of hay along the cart track, cutting the bailer twine and throwing sections of hay over the dry stone walls to the cold, hungry animals.

Conrad went out one morning, taking hay to the top fields. The snow was already very thick and treacherous under foot. The wind was blowing, it seemed like a force 10 and he was staggering about with a bale in his arms. Suddenly he missed his footing and went down, wrenching his knee in the process, very badly. Unable to get up, he lay in the snow for ages and by the time he had managed to crawl back to the house and we could examine him, his left knee had already swollen up to twice it's size. Heavily bandaged, he continued to work on it during the ensuing weeks, in agony. Unfortunately this set up arthritis for the future years and eventually resulted in him having to have both his knees totally replaced with artificial ones.

Obviously I helped all I could but even so it was a hard, exhausting time. Sometimes the winter seemed to last an eternity, the darkness coming down and enveloping us in a black cloak by about 3.30 in the afternoon. The icy winds moaned day and night and power cuts were frequent. Fortunately we did have a good supply of oil lamps and candles and always, at least, the comfort of a crackling log fire and a glass or two of sloe gin!

CHAPTER 11

I have often expressed a desire to ride again and now, with all the acreage we had, it seemed natural enough that my thoughts turned to having a horse. We had already cut our teeth on a large, soulful donkey, aptly named Jenny, whom we had bought from a dealer in our first heady months at the farm. Jenny arrived and we were assured she was already in foal again, with another at her foot. A tiny elegant little beige filly, whom we called Bella. She had neat shiny hooves that clicked like a pair of castanets every time she walked and ears like small kites. Jenny was lovely but did nothing at all. She was more like a portable lawn mower. The effort of moving from one luscious clump of thistles to another was just about all she could manage but if I sat Katie on her back, the head came up abruptly, the bat ears would flatten and the feet firmly dig in. She would move neither forward or backward and if a carrot was proffered for bribery, she would stretch her neck just far enough to grab it. Thwacking with a stick produced no effect whatsoever, neither did any form of sweet cajoling. She gave the air of being permanently depressed, as if the cares of the world were on her withers. This, plus the fact that she was slightly flat footed due to a previous bout of laminitis, made her appear to be the most down-trodden

93

creature on God's earth.

The idea of actually using her was totally out of the question, and the visions that I had harboured of sunny days, with straw hat perched on head, to round up cattle or, for that matter, the odd small trip to the village to gather groceries, perhaps with a small painted trap - soon evaporated. Jenny was extremely good at one thing only and that was eating.

After our London house was sold, browsing through the local paper, I happened to see an advertisement - FOR SALE - LOVELY GREY MARE, 4 YEAR OLD, PART THOROUGH BRED, 15 HANDS. GENTLE NATURE. APPLY; THE MANSE.

I pointed it out to Conrad who, bless him, was prepared to indulge me and we rang the number. The resident of the Manse over the valley, was a vicar and his wife and he assured me that he was indeed riding the horse every day. He used her to visit his parishioners, tying her to their fences as he went. I must say, visions of a man of the cloth galloping around the countryside in true Western fashion to comfort the needy, seemed more than a little strange but this was Scotland and anything was possible.

We made arrangements to come and view the horse and I took a girl friend of mine along with me, who kew her horseflesh. Avril was a small, petite lady, with a shock of bright red curls. She lived on a large neat farm some miles distant and horses were her passion in life. So much so, that I was nearly led to disaster because of it.....

"Come on over," she had said one morning. I had foolishly told her that I had ridden the odd dobbin in my childhood and had a hankering to have a gentle ride again.

"I'll saddle Marcus, he's very gentle," she said.

I knew she had about four horses at that time and stressed that I had not been on the back of a horse for about fifteen years.

I arrived round at her farm, full of a mixture of anticipation and mild apprehension.

We went round the back of her house into the yard. Blowing steam and snorting around, already tacked up, was the largest black horse I have ever seen. He seemed to go on forever. I looked quickly at the other horse, a pale chestnut, who looked equally large but far less alarming and I inwardly prayed that he was Marcus.

"We'll just go for a gentle trot through the village and head back home through the fields, to start with. Just to get you a bit used to being in the saddle again. Nothing too strenuous to start with," Avril said. I nodded mutely. What had I let myself in for? With practised ease, Avril mounted the chestnut horse and waited patiently while I tried to get my foot into the stirrup of the large black monster, who was obviously anxious to be off. Finally, in a burst of giggles, Avril dismounted and gave me a leg up.

"Are you sure he's all right?" I asked rather tremulously.

"He's a lamb - just follow me," she said gayly.

We walked out of the yard. I was very unsure of myself. I felt as if I had an elephant between my legs not a horse. My hands were uncoordinated and I felt as if I was on a breathing time bomb. The first stage of the ride was really quite pleasant. We walked sedately along narrow country lanes, bathed in the most glorious sunlight. The world looked quite different from up there. Magic! I began to relax. We trotted (in a rather hit and miss fashion) through the village, the mountain of black flesh powerful and sure. Oh this was the life. My trotting was a bit ropey true but given time I was sure it would all come back to me...! It's amazing how dumb you can be on occasions.

Lulled into a false sense of security, I waited while Avril opened a gate into a huge meadow that seemed to and indeed did, stretch for miles. No sooner had she closed the gate, than the latent time bomb ticking away beneath me, exploded. With a lunge, Marcus tossed his head, snorted a few times and took off. I have never been so terrified in my life.

"Whoa" I yelled, tugging futilely at the reins. Marcus had different ideas. We were now at a flat out gallop. Tussocks of grass came up to meet me. I pulled, I yelled and above all, I hung on. Several times my feet came out of the stirrups altogether and at one stage, managing to fling a glance behind me, I saw Avril about 200 yards away, gently cantering on, splitting her sides with laughter. I must, from the rear, have looked a sight for sore eyes. Nothing I did at all made the slightest impression on Marcus. We galloped up hills, over seemingly hazardous drops, through mini-ponds, round fences, over tree stumps and under low hanging branches. Several times I prepared to meet my Maker and several times I almost did. After what seemed an interminable amount of time, Marcus finally dropped speed and just simmered to a very fast trot. Avril caught up with me, still in stitches. She opened another gate, which led into the rear of her yard and Marcus finally skidded to an abrupt halt inside.

I slid off his back, my legs like jelly.

"Jesus, Jennie," Avril said. "I'm sorry. I forgot to tell you that I usually give him his head when we come through the fields, because he knows he gets fed when we come back".

"Thanks for telling me," I muttered.

"Well, it's been great," she said. "You must come up again sometime. I know some lovely rides around here".

"Yes," I lied. "That would be nice". A still, small voice inside me completely disagreed with me. Besides, I felt I'd done all the lovely rides in one potted version.

With difficulty I walked to the van. It was not until I actually got into the van, that I realised my legs would not meet. I had to physically put one foot on the accelerator and one on the clutch. God help me, I thought, if I need to use the brakes!

Back at home, I got out of the van and walked down the garden path to find Conrad. He looked up from digging.

"How did it go?" he asked. Then "What have you done?" I was walking as if a herd of Buffalo had passed between my legs, completely bandy.

"Oh it went fine," I lied. "I just bolted on a big black stallion called Marcus, rode for about 20 miles at a flat out gallop and feel as if every bone in my body is broken, apart from that it went fine!"

"Jennie, you do exaggerate" he said.

You would think that this little episode would turn me off horses for the rest of my life. Not so. Hence, armed with Avril, we drove over one summer evening to look at the grey mare belonging to the Minister.

At first sight, he and his wife were an odd couple. He was extremely tall and slim, whilst his wife was extremely small and quite dainty, with quite a strong voice. He was soft spoken and ethereal. Two noisy children screeched about the place. We followed them over to the paddock and there, munching away, was the grey mare - Sophie.

I knew as soon as I saw her, that she was just what I wanted and I started expressing so to the Minister.

"Listen," hissed Avril, "don't be so darned quick. We should see if she's all right to ride first. She might be terrible".

"Would you like to try her out?" beamed the Minister. Not wanting to make another spectacle of myself with everyone looking on, I declined the offer.

"I'll just give her a wee go," Avril said firmly and I left her to it. She was fine and the deal was struck. We arranged to pick her up in a horse box of Avril's, a day or two later. My elation knew no bounds.

Sophie soon settled in to her new surroundings and I did in fact have several gentle saunters into the world with her but she was not very fond of the sound of running water and as soon as we got anywhere near the burn that bubbled along the bottom of our lane, she showed her displeasure by walking backwards up the side of a grass verge, rather than cross the road to get past it. Basically though, all in all, she was extremely placid and I was anxious to have a foal from her.

On one lamentable occasion when I had been for a ride on her, just to the village and back, we were ambling along a country lane. Sophie gazed around her in mild curiosity, the birds twittered and we passed a nice family sitting having a picnic by the road side. They smiled and waved and I waved back. We were doing a sedentary walk - it was a pastoral scene. Just as we passed the picnickers, Sophie - not looking at what she was doing but probably at an egg sandwich - tripped up. I was of course totally unprepared and smartly fell straight over her right shoulder, landing with an ungainly thump onto the tarmac road. Spilling their sausage rolls and bottles of Pepsi around them, the family sprang to their feet and asked if I was all right. Boy, did I feel foolish. Sophie got to her feet in a rather unladylike scrabble and stood looking at the sky, trying desperately to disassociate herself from us all. Assuring them I was completely unharmed, I mounted my trusty steed and plodded back home. I mean to say, honestly, to fall off walking, after hanging on through the perils of Marcus, was just too much!

"The best person you can go to if you want a good foal, is Andy Spense," said Avril. "He's out Corrilaw way and has a beautiful stallion called Hesonite. Why don't we go along and have a look?"

Corrilaw was another place out of 'The Big Country' stretching for mile upon mile of breathtaking scenery. Andy Spense's farm was a beautiful old building, surrounded by stables and barns and fields. Most of the nearby fields contained brood mares and foals, with a few yearlings. The whole place was extremely well cared for and presented. Hesonite was indeed all that we had imagined and more. Seventeen hands of liver chestnut, looking almost bronze in the light, with the longest, waviest mane I had ever seen. He looked so perfect, like something out of a Stubbs painting. I felt pretty sure that if I liked him, Sophie would, and we arranged to bring her over to be covered within the next few weeks when she was due to come into season. Andy Spense was a charming man, who obviously knew his horse flesh. Hesonite was an ex-Derby winner and had a reputation for siring some extremely fine youngsters.

"No foal - no fee," Andy said. "Bring your mare down and let her stay for a week. You pay me the fee if she produces you a foal". And so we did... and so she did!

May was a wonderful month at Skeoch. Everything starting afresh, the harsh winter months at their end (although frosts were still likely to occur) and the fields and hills covered in spring flowers. The new bracken on Skeoch Mount just starting to push through in delicate pale green fronds and always the belief of a fresh beginning, with the summer ahead to look forward to.

Sophie's belly had reached a good pear shape, the veins distended. She found pregnancy now rather irksome. I obviously no longer rode her and motherhood was imminent. Every night before going to bed we would go out to see how she was. I had put her in a small paddock by the side of the house (called the field of sinks, because there were a number of them thoughtfully put about the place to catch the rainwater). It was secluded and protected. Night after night we went out and night after night there was no change.

One evening some friends called in for supper. We had a jolly time and the night wore on with plenty to eat and drink. It got later and later, with our friends reluctantly bidding us good night at about 2.00 a.m. Feeling rather tired we decided not to go out and look at Sophie. We had kept an eye on her throughout the day and there seemed no sign of her 'bagging up' - of milk coming into her teats. We went to bed. I slept fitfully and finally woke at about 5.00 a.m. I slid out of bed, donned wellington boots and dressing gown and went out. There had been a late frost that night and in the half morning light I shivered as I walked to Sophie's paddock. She was standing up eating, looking exactly the same and totally ignored me. As I turned to go back to the house, she suddenly whinnied and, as I turned, her waters broke and the first stages of labour started. I leaped back to the house and yelled at Conrad. "Get up, get up, she's started labour". Con, poor soul, leapt out of bed like a scalded cat and together we quietly walked to the paddock. Sophie was by this time, lying down, tiny front hooves were trying to emerge from her heaving body. It is said that horses like to be very private when giving birth and they also have to give birth fairly quickly, unlike cows, who can seem to go on in labour for hours without any ill effects. Conrad and I crept along the side of the fence by the paddock, in order to keep a watchful eye, without disturbing her. After about an hour, the frost still very much in evidence, we decided we should go indoors have a cup of coffee and to be on the safe side, call the vet. Labour hadn't seemed to have progressed very far as yet, despite a good deal of pushing and grunting from Sophie.

The vet wasn't exactly overjoyed at being hauled up at 6.00 a.m. on a frosty morning, but assured us he would be up fairly soon. We paced the floor like a pair of expectant fathers and decided to go back out and keep watch.

There are some moments in life which are totally unforgettable. This most certainly was one of them because just as we arrived at the paddock, the sun flooded out from behind the grey clouds, turning the grass into a maze of dazzling crystal. Sophie neighed and pushed and out plopped a large, wriggling sack. A tiny chestnut head came up and a shrill little whinny filled the air. The foal, still parcelled up in his 'polythene bag' struggled to get to his feet. Sophie nuzzled him, his wet head, shining and steaming in the sun. He was a bright copper gold chestnut with a white flash on his face and three white socks. His legs were quite the longest things ever and totally at odds with the rest of his body. Like some unsure ice-skater, the legs took on an independent life force of their own. For at least half an hour, we silently watched an incredible ballet. The vet arrived up to give a tetanus injection to Orlando as he had been christened (after Conrad's first Shakespearean part on stage), and ordered Conrad to hold him firm. Now Sophie was normally the epitome of good behaviour, but seeing her newly born offspring being manhandled was too much. She promptly laid back her ears and bit Conrad smartly on the arm. The air was a bit blue! However, all was now completed and we had witnessed an unforgettable scene. Events like these made some of the harder times more bearable and helped to weave a web of magic, that seemed to bind us to the place.

Now that we had decided, or rather, had thrust upon us, some sort of farming, it became increasingly clear that we would have to consider some sort of animal fodder for the winter months - and this among other things, meant hay. It was incredibly expensive to buy in and the quality was not always good. We had no implements like a bailer, hay cart or turner. A turner was in fact called a 'wuffler'. I don't know why, perhaps because it 'wuffles' up the hay. It is a key piece of equipment for haymaking and looks rather like two giant egg whisks, which are attached to the back of a tractor. Within minutes it fluffs up the hay, spins it round and leaves it lying in neat, dry rows, ready eventually for the bailer to bail up. We would look longingly at the farmers in the fields who possessed this piece of machinery. It could turn the hay in a large field within a few hours. However, we didn't have one, so we had no choice but to turn the hay by hand ourselves, with a couple of strong pitchforks.

We had totally unrealistic visions of help, perhaps from other local farmers, pitching in to lend a hand. I visualised a scene, like something from one of Constables paintings, with picnic lunches spread under the trees (on gingham cloths!), while the men raked and packed the hay into the patiently waiting horse cart. Dream on! At that time of the year, the other farmers were busy making their own hay, with reasonably up-to-date machinery. Still, not daunted and certainly not knowing what we were letting ourselves in for, we arranged to have the grass cut in the hay field, as soon as it was ready.

We felt enormously excited when the contractor arrived up at Skeoch with his powerful cutting machinery, one fine sunny morning. Thank goodness we didn't decide to scythe it as well.

He worked hard all day and by early evening Conrad and I roamed across the heavy swards of grass, breathing in the heady smell of new mown hay. It was glorious and we were anxious to start turning it as soon as it had dried out. We decided to make it into ricks, the old fashioned way, with a tripod of sticks tied together at the top. The dried hay is then carefully packed round and round, until a pyramid is obtained at the top. Before any of this could be achieved, of course, the hay had to be turned by pitchfork, over a period of several days or weeks, depending on if it rained, until it was completely dry. Wet hay packed down, goes mouldy very fast and can spread mildewed spores through the dry stuff, wasting it all.

Apart from the sheer physical hard work of turning it hour after hour, in the blazing sun, eaten alive by the 'Clegs' which were a giant form of horse fly that could bite through the toughest cotton shirt, the frustration of seeing it cloud over and gently pour on your last eight hours sweaty effort, was almost too much to bear. We had to wait until it had dried out completely on the surface again, before turning it once more. On a ten acre field it amounted to about fifteen tons of hay. Oh the blisters! Oh the arm ache, the bites - and of course, as sometimes the case, given the nature of Conrad's profession, in the middle of all this - he was offered work back down South again. Heidi for the BBC. Conrad was torn. The work offer was very good but we were desperately trying to get the hay together. We had had too much rain and were running behind schedule. The long range weather forecast was looking good. Conceivably we could do it but not if Conrad was away. There was in the end, no option. He had to take the job - that ugly word finance dictated. I would stay at Skeoch with the children and we would somehow have to seek outside help.

It was a very concerned Conrad that finally said goodbye a few days later in the yard. He had worked like a demon right up to the afternoon of his departure and was heading for the long drive back with no energy reserves at all. Not the greatest of conditions to be in for rehearsals, I thought. I was more than a little apprehensive myself when I waved the white Renault van out of sight.

I worked in the field for most of the next day, having managed to send the children for the day to my friend Cathy, who had two children of her own. I broke off work only to feed the livestock and myself. Later, when I collected the children and gave them their supper, bath, story and bed, I waited until they were asleep and went back out into the field, with the child alarm strapped to my waist. I worked on and by around 10.00 p.m. the dew was beginning to fall, although it was still very warm and light. I decided to call it a day. Wearily I trudged back up the field, aching in just about every limb, my hands pretty blistered. I looked up at the turquoise sky, the softer shades of lilac and deeper purple night clouds just brushing long, soft fingers across the horizon. A sliver of new moon, pale and iridescent, like a baby's fingernail, hung in the air. I stopped and stared at it for a few minutes - and made a wish.

I was moaning away to Avril on the telephone a couple of afternoons later. Telling her how exhausting it had all become.

"You should get in touch with DK," she said.

"Who is he?" I asked.

"Davey Kirkpatrick - but he's just known as DK. He's got all manner of farm machinery and he may be able to help. I'll give you his number".

Later I spoke to Conrad and he readily agreed that I should give this DK a call and see if there was a possibility he could help us. I wasted no time in ringing him and he kindly agreed to come up the following day - with a bailer.

DK was a dealer. He dealt in anything and everything. Horses, calves, sheep and farm implements. He was a Mr Fixit. You want it, DK could get it. We were to visit his farm on many occasions in the future. It really was like something out of The Darling Buds of May. Piles of broken bits of machinery lying in the yard. Goats and chickens roaming free range. Nothing seemed to be thrown away. Everything came in useful. Corners full of partially stripped down tractors, wheels, gear boxes and oil drums.

Margaret was his young wife, who clearly adored him. She kept their little farmhouse spotless and was the essence of femininity - right down to her 'dolly' knitted toilet roll holder!

DK himself had a kind, weather-beaten, slightly

battered sort of face and a very easygoing manner. Nothing seemed to daunt him and he had a cheerful, optimistic attitude to life. His father-in-law George, also came up with him in the afternoon to bale. He was a stout little fellow with a cheery disposition. They arrived up as good as their word, with a huge bailer and set to in grand style. It was wonderful to see them fork the hay into the bailer and neat, string tied bales drop from the end. They fell about laughing when they managed to get fourteen bales out of one of our ricks. (Fourteen bales is a LOT of loose hay!) "Ach" said father-in-law, spluttering, "tha's nay a bloody rick, it's a wee stack more like". Out of about fifteen tons of hay, they finally baled about ten and left the other five tons of loose hay for us to get under cover ourselves. That was marvellous and I felt relaxed and happy that we could probably manage that.

Hay is an amazing thing to try and control, especially on a "no sided" hay cart, which happily DK had come to the rescue yet again, with. The children had great fun leaping on top of each precious load and we patiently packed forkful after forkful on top. Thank goodness Conrad was now back from his work down South and we both set to work with a will.

Along with the hay cart and flushed with a 'repeat fee' for Conrad, which had unexpectedly come in, we also bought an extremely ancient, red tractor, which was immediately christened Hannibal. (Don't ask me why!) It had no roll bar or cabin and appeared to have no brakes as well. None that worked anyway. We did have a clutch and accelerator though, so when you wanted to stop you stuck your foot really hard on the clutch and jammed your other foot as hard as possible on the middle peddle - and prayed. Sometimes it worked... but not always. Trying to stop on a 45 degree angle with a full hay cart behind you, was hazardous to say the least.

We wobbled our uneasy way bit by bit to our rather decaying outside cattle shed. The hinges were rather dodgy in the door frame and the corrugated iron roof was not very stable but it was some kind of shelter. Once there at the shed, the whole lot of hay had to be unloaded by pitchfork into the shed and back again for another load from the field. The wind usually managed to catch you, just when you were manoeuvring with a tricky bit, scattering it back over the entire field. It was a frustrating business and still took far longer than we anticipated. After some weeks of near total collapse, we finally had it all under cover. The great old cattle shed door shut and bolted. We felt good. We felt very good!

To celebrate our "Harvest Home", we decided one morning, a few weeks later, to pop to our local, which was about eight miles away at Dunscore. It was a Saturday morning and rather blustery but we were in high spirits. We had at last got on top of the work factor and were beginning to see our way clear to having a relatively easy few weeks. Throwing a joint of pork in the oven on a very low setting, we put the girls into the back of the van and started on our way to the pub. It was a mini-Pyrenean journey, going the picturesque way, round the most hair-raising narrow hair-pin bends, through umpteen gates and cattle grids. The sun broke through the scudding clouds, dappling the gold and green hillsides. Stubble fields glowed from Sepia to Bronze. Flocks of sheep huddled by the spinneys and brooks, impervious to the wind which had by now picked up considerably, their backs turned firmly to the world. I felt that those late summer days were the best I had ever known in my life in terms of their beauty and the incredible feeling of being "at one" with nature.

We passed a jolly hour or two at The George, catching up on local gossip with some of our casual friends. Glancing out of the pub window however, we saw the odd small branch whirl past. One of the locals turned on his bar stool and muttered. "Aye, we're in for it - noo doot. We allas get it aboot noo!" He stared reflectively into his pint of 'heavy'. Con shot me a glance. It was beginning to look decidedly grim out there. The sky had darkened considerably, rain was definitely on the wind, and the wind itself was more than just gusting.

"I think we'd better be heading back, Jen" he said. "I don't like the look of this. The wind is positively howling". We dashed to the van, rather alarmed at the strength of the wind and looking forward to getting back home and snugged up. Once in the van however, we put on a Glen Campbell tape and sang all the way back. Skidding our way back up our cart track, over the pot holes, laughter and song still in the air. I was looking forward to a nice roast dinner, with perhaps a glass or two of bramble wine and a good roaring fire. Above all was our enormous sense of relief that we had at last managed to get the hay stacked away for the winter.

Our voices cut short in unison at the sight of what confronted us.....

Blocking the entire top of the lane, were sheets and sheets of corrugated iron, some still bolted together in a twisted, mangled blockade. Some had wrapped itself cosily around the telephone pole, whilst other pieces waved gayly in the wind. Most of the outside cattle shed which had housed our loose hay, so carefully and lovingly packed in, had succumbed to the wind. The roof rafters were gaping open to the sky and our hay - that precious hay, spiralled its way heavenwards and beyond. I didn't feel at all "at one" with nature now I can tell you.

Conrad and I got out of the car and pulled some of the iron sheets to the side of the lane, so we could get the van through. I cannot begin to describe our feelings at that moment. What little hay that remained within the three standing walls of the cow shed, was ruined with the deluge of rain. It was soaked. The shed door had been torn off its hinges and tossed into the field. The rest of the hay had blown itself just about everywhere imaginable. All that work and effort - for what?

We went into the house and had lunch. There was nothing we could do until the wind had abated and then we would have to start the cleaning up process. It was abundantly clear that we needed more machinery if we were to make any success at all at farming. That meant more money ...and so on. We took some small comfort that at least we had ten ton of hay bailed and safely stored in the Dutch Barn.

It taught us a great lesson and for the next summer we bought an old "wuffler" from DK ... and a bailer!

CHAPTER 12

....''The fault dear Brutus is not in the stars...''

No, the fault I suppose, was ours in not watching out for the fact that we had had precious little rain for about eight weeks and the reservoir for water was now dangerously low. The pump in the top fields had spluttered a few times and finally, noisily, breathed its last. A muddy trickle came through the kitchen tap. Talk about all hands to the pump! Armed with spanner, water funnel, saucepans of brackish water that we had gathered from the catchment well near the house, we fled to the top fields. The pump was airlocked; a few inches of rusty water stood in the large water trough by the reservoir and about thirty head of angry Angus and Hereford cattle were nudging each other for the last dregs. Racing the six hundred yards to the well, we dragged the lid off and peered in. It had, in fact, half filled but the pump had sucked in air when the water was too low and was now not pumping up to the reservoir. Priming the pump was an arduous job at the best of times. One had to turn the electricity supply off, which had thoughtfully been placed in the loft of one of the cattle byres, back near the house. Once the supply was off, the top of the pump had to be removed, a funnel placed over the pipe end, water poured in, in a continuous motion and the electricity supply turned back on again...(in the loft of the cattle byre!) Sometimes, rarely, you were lucky and it 'caught' first time. More often than not, it just guzzled all the water - and still remained airlocked. After racing back and forth about six times, on this particular occasion, we finally got it started. I ran up to the reservoir and left Con still frantically pouring water down the hungry mouth of the pump. The cattle by this time, could hear water trickling into the tank but couldn't get at it because of the overflow device attached to the reservoir, which in turn - fed the cattle trough. It would be some time before there was enough water in the reservoir to pipe through to the trough. The heat was searing and the flies and clegs hung in a black cloud around our heads, making the cattle and us, even more angry and frustrated.

Sweatily Conrad panted up to me. "There's nothing for it," he said, "except to bale out the water as it comes in and pour it manually into the trough".

Fighting our way through the cows, we pushed aside the massive lid that covered the reservoir. It must have weighed about half a ton. Then we started. Bucket after bucket. As soon as the first was poured into the trough, it was quaffed in what appeared to be one long gulp, by thirsty cows. The biggest and fittest elbowed the smaller ones out of the way. Each one took at least three bucketfuls before grudgingly allowing another in its place. Hour after back-breaking hour we toiled, our hands blistered and our backs aching like hell! The sun was going down by the time the last thirsty cow had wandered off to eat grass. Wearily we went indoors for a much needed cup of tea. It would be some time before we could afford ourselves the luxury of a hot bath or the pleasure of being able to flush the loo! Every hour we had to climb up in the byre loft, switch off the pump to prevent it sucking air and allow the well to gradually fill, before climbing back up the loft again to switch it back on. Oh the appliance of science - the marvels of electricity!

Katie by now had reached school age. In the local village some five miles away, was a very pretty little school building with three classrooms, assembly hall and homely kitchen. It was run, very efficiently, by a lovely lady Nan Kellar, who always managed to look like someone from Vogue magazine - Scottish style. She was tallish, had a beautiful lilt in her voice and pronounced words like honey pouring from a jar. Altogether there were three teachers for the three classrooms; thirty pupils altogether. Wonderful - ten to a class! All the local village children went there and the standard of education was extremely high. Katie blossomed and it wasn't very long before her little voice took on the same lilt as the other wee bodies.

"Wee Wallie Wenkie, roons thu the toon, Upshtairs, doonshtairs - in his nichtgoon" she would chant. Every so often I would bring home a Fiona or Kirsty to play and long merry hours were spent, swinging from the Tarzan rope in the Dutch Barn, off the bales of hay, or trying to ride a disgruntled black Shetland pony, bought in a fit of weakness and called, of all names - Prince! Needless to say, Prince had other ideas on the subject. Like a true Thelwell cartoon pony, he would roll his eyes, wait until a little body was just about poised to straddle a leg over and then he would move smartly forward a few paces, leaving some small soul in a heap on the grass. He had honed this technique down to a fine art. If really hassled, he was not above planting a neat black hoof sharply behind the back of one's knees, which had a most pleasing pole-axing effect upon the unsuspecting victim.

I once watched Katie, in the tiny paddock by the side of the house, trying to mount Prince. He was tacked up and she patiently led him to a large rock which was to be used as a mounting block. Sweetly he stood still, accepted the proffered carrot, waited until she was virtually airborne and then just moved forward enough, for her to land, face first in the grass. I think I counted this performance no less than six times, but undeterred she finally managed it, not without getting rather cross in the process though! However, her patience was rewarded. Getting him to walk however, was another matter.

Prince was bought off a dealer for £45. His age was indeterminate but I was assured at the time that he was well broken and "a little dream to handle". When he started shedding his teeth, I started having grave doubts and one day, whilst the vet was up to inoculate some cows, I asked him to have a look at the pony's teeth and give me some idea of his age. Prince did not like being manhandled and tried to put up a fight but the vet was well used to the likes of such small fry and finally prised his jaws open.

"And how old was he sold to you as?" the vet queried.

"Well, I wasn't really given his age but I was assured he was used to being handled. I had the impression of some age," I replied.

"He certainly is," the vet grinned. "All of three. Those aren't his full teeth he's losing, they're his milk teeth," he chuckled. "No wonder he's so stubborn - he's nought but a bebbe!"

When we purchased Skeoch, along with the pastureland, we were fortunate enough to buy the two and a half acres of woodland, in which, in the first wild enthusiastic days, Conrad would spend many hours, carving the little stream that flowed through, round boulders and making waterfalls. In the warmer months it was bliss to wander down there and gather cuckoo flowers, marsh marigolds, ripe grasses, along with tangles of wild sweet peas in delicate shades of lilac and pink. The children would run down there. It was their magic place and Katie would trundle little Sarah along. Together they would sit for hours "talking to the Fairies". Their special bond grew in the shared happiness of new discoveries, new delights and almost complete freedom to roam in the most delightful surroundings. Talking to the fairies was not always enough however.

The daily chores for me then, apart from helping to run the farm, were time consuming in their own right and my days were well divided between being a wife, a mother, first mate, part-time vet and farmer.

One glorious summer Conrad and I had spent a lot of our time outside. There was the hay to be cut, walls to be mended and a thousand and one things to attend to. Kate was by now five years old and Sarah was two. In truth, we had not devoted as much of our time to the children in the last two weeks as we should have liked. Obviously it was time, according to Kate, to be taught a lesson.

Conrad and I were climbing up from the far fields, towards the house. There, emerging from the bottom of the garden, heading towards the woods, we saw to small figures, hand in hand. Over her shoulder Katie had a stick, with one of my headscarves tied to it, Dick Whittington fashion. When we finally caught up with them, two little faces stared intently in front of them.

"Hello girls," we smiled, "where are you off to?"

With the utmost seriousness, Katie's earnest little face studied us.

"We are leaving home, aren't we, Sarah?" she said. In her hands was a book of Homer's Odyssey and a book of Greek Tragedies. In the parcel on the stick was a packet of rich tea biscuits. Sarah looked up at Kate and nodded, trying not to giggle. In her hands she carried a bottle of water, nearly as big as herself.

"We are leaving home and going to live in the woods with the fairies," she said gravely.

Obviously this had to be given some serious thought and stifling the urge to laugh at the sight of such a determined pair of little people, I immediately felt a strong pang of remorse. Some serious gestures would have to be made to rectify the situation. We asked them, very kindly, if perhaps they would reconsider their rather hasty decision. Special cake at tea-time, longer in the bath, a bit more television to watch (which was quite heavily monitored), extra stories at bed time. Bargains were struck - Katie was no fool. Promises, bribery, cuddles, lots of love and a lesson learned. Work can take you over. That incident acted as a gentle reminder of what happens when the scales are tipped too far one way.

We woke again, to a peerless blue sky. It was mid summer. The heat, that had been building up for weeks, already at 8.30 a.m. was forming a watery haze, bouncing and shimmering off the whitewashed walls of the byres and radiating back on itself from the concrete yard. The air crackled with electricity. You could smell it. This particular summer was the hottest then on record, for some considerable time. Cattle stood around under the trees, getting shade where they could, already angry, stamping their feet and tossing their heads, while clouds of black flies plagued their eyes and legs. Nothing wanted to move. Every effort produced a film of sweat. My T-shirt stuck to my back. It was too hot for the children to play outside and, like everything else, they were fractious and quarrelsome. The ground was parched and we used our water meagrely, worried in case the well should again run dry. The cattle had to be our first concern. Never before had we longed for rain, anxiously searching the sky each day for signs of a let-up. Storm clouds gathered occasionally, raising our hopes, only to dash them again a few hours later as they melted into the scorching skies. Thunder had threatened for days but never actually happened. It grumbled intermittently, like us. As the storm clouds gathered, so our tempers got just a bit more frayed. Mundane jobs like fence-mending or mucking out the byres, seemed more tiring. The flies were a constant irritant, settling on you the moment you stepped foot outside. The heat was sapping, uncomfortable and sticky. The only respite from the heat was either the middle of the night or the early hours of the morning, neither very practical from the farming point of view. The bad-tempered cattle were now having to eat our precious hay, normally reserved for the winter but the grass was parched and brown. There was nothing substantial left on it. In desperation, one afternoon, we all went down to the woods by the bottom field. The little stream had completely dried up and the natural pond reduced to a mud hole but the children were at last playing happily, under the shade of the trees as Conrad and I gathered up enough energy to mend some fencing.

After a few hours, we started to trudge wearily back up the field towards the house. The sky had darkened and the wind had picked up but it was a hot wind and we had been fooled before. Then - it started. In fat drops the precious rain slowly fell. You could almost hear the grass sigh with relief. The pace increased, the drops became a torrent. We shouted and we yelled in delight and danced about in it, getting absolutely soaked. The children promptly ripped off all their clothes and ran around, their little brown bodies cavorting about the field, like a pair of water sprites. Oh the relief and joy. It poured and poured. I have never been so happy to see the rain. By early evening, the first of many deluges had stopped. We went outside to a perfect sunset. Everything was pristine and glistening. Life had already come back in the garden. The cattle, still steaming from the soaking they had had, looked cool and content. Some of the water still lay in large puddles in the fields. The sun-baked earth couldn't cope with such a large quantity of rain. The chickens wandered about the yard, scrawny and bedraggled, their feathers plastered to their skinny bodies. There was copious shaking and ruffling of feathers and a lot of chatter and squawking. The air smelled wonderful. Personal harmony was restored.

Later, when we went back indoors, we all allowed ourselves the luxury of a really deep bath and lying there, wallowing up to my neck in the sweetly smelling foam, I thought how easily harmony had been restored to everything by that first deluge of long-awaited rain and pondered on how much we take nature for granted. Having only minor drought conditions to cope with, compared to some countries, was enough to make us all irritable and distressed. Imagine living in Somalia or Ethiopia. Living in the country, there is such immediacy when the sun shines. You can be in it straight away, not longingly looking out of an office window, hoping to grab a few hours when you get home, if you're lucky. The drawbacks equally exist of course, when you have to trudge through the snow, hail and rain, whether you want to or not.

From a child's point of view, Skeoch was a paradise. A totally natural environment, learning about life and death on a day to day basis. The children watched animals being born, chicks hatching and also the ultimate demise of things. It instilled in them a deep love of the countryside and respect for nature, which has happily stayed with them. When a child can witness first hand the miracle of birth, it is much easier to explain about so many other things, which by their nature, sometimes seem difficult or unpalatable. There was an easiness in explaining the facts of life. It was no big deal; all things were explained and understood when the time was right. They had a natural curiosity and everything was discussed openly, whether it was the size of Orlando's willie or why the cock kept jumping on the hens. You can't keep questions at bay forever!

Kate and Sarah's growth and development was a wonderful thing and although the farm was hard work, I was lucky to be able to share and be with my children at that time. As a person I was totally fulfilled in my life, with my family and the farm.

CHAPTER 13

The constantly changing moods of the countryside around Skeoch never ceased to amaze and thrill me. There were dark days when the whole of Skeoch hill was shrouded in a heavy mist, brooding silently away, at times rather threatening. You could almost hear the voices of the Covenanters, echoing out through the vapour, their secrets still safely hidden among the bracken and gorse-covered hillside. In the late summer, there was an eerie light which used to fall in the early evening, cloaking everything in ethereal beauty, for Skeoch Mount was more than just an ordinary beauty spot. It's very roots were steeped in a much greater and bloodier history.

Charles II of England considered that "Presbytery was not a religion for Gentlemen". He had signed only two Covenants in 1649 in order to secure his own coronation. He had absolutely no intention of carrying out the religious and political obligations he had taken on, or those undertaken by his father before him. His Officers of State and Privy Council he chose without any reference to Parliament. The Scottish Parliament at that time in 1661 was packed with Royalists who basically rescinded most of the measures passed since 1633. This meant, in effect, the restoration of James VI's method of himself choosing the Committee of Articles, which strengthened the King's position in relation to Parliament. As far as the Church was concerned, it had the effect of bringing back Bishops and restoring the former system of patronage, which was that Ministers were chosen by the Laird and not the congregation.

James Sharp, a covenanting minister, was sent to London from Scotland in order to make representation at Court as the emissary of the Presbyterians. He returned, some time later as the Archbishop of St. Andrews and, of course, now expressed entirely different opinions than those he had set out with! Now new regulations demanded that ministers appointed since 1649 were required to resign their posts and receive them again from their Bishops and patrons. The majority of ministers reluctantly agreed but some 300 or so refused and left their manses and churches rather than submit. What is more, they had, particularly in the South West, the support of their parishioners, who were very steadfast. Before too long unauthorised religious services were being held, in secret, by these "ousted" ministers. In houses, in barns and even the bare hill sides. One of which, of course, was Skeoch. Troops were eventually sent in to collect "fines" from those that attended the 'illegal' conventicles and armed clashes followed. This only served to endorse the loyalty of the Covenanters, who were prepared to die for their cause if necessary and in spite of terrible punishments inflicted on them, their resistance continued.

Archbishop Sharp was finally killed in 1679 by the Covenanters and the government, in consequence, cracked down even harder. Troops were sent in and 1400 prisoners were rounded up and sent to Greyfriars Church in Edinburgh, where they were all murdered. It is not surprising that the 1680's were known as the "Killing

Time".

Now our local folklore told of the Covenanters holding their secret meetings somewhere near the top of Skeoch Mount. Claverhouse and his men had rampaged over the whole area and rumour had it, that at Jim Kissock's farm, The Scour, at the bottom of the lane, one such brave fellow was finally and deliberately starved to death in one of the outhouses, by Claverhouse's men. Whether it was true or romantic fiction I don't know but it certainly wasn't beyond the realms of possibility.

However, every year, whilst we were at Skeoch, people from all over the place came on pilgrimages to try and find the exact place where the prayers and secret services had been held. Many asked us directions but it wasn't something you could direct anyone to. It was far too complicated for that. If it had been that easy, Claverhouse's army would have found it...

The path up the mount was quite tortuous through the bracken and brambles. The hill itself was very deceiving because just when you thought you must surely be at the top, there was yet another hidden valley and yet another series of small hills.

Conrad and I made the journey one day and started out brightly enough. After about one and a half hours' climbing, we stopped for a break and looked back down the way we had come. Nestling in the hollow, like a dinky toy farm, was our little Skeoch farm. The only building anywhere in sight. I thought then, of how many pairs of eyes had seen that same scene as us and how little things had changed in all that time. After a bit more climbing and just when we were going to call a halt, quite unexpectedly, we came upon a most beautiful glade. We were there - we had found it. In the middle of the glade a large circle had

121

been formed, where forty or more flat, granite stone slabs of huge proportions had been carefully laid. In the very centre of the circle, on a rock plinth, stood a carved stone chalice, covered in pale green lichen moss. It towered some six foot high and was awe inspiring. The atmosphere there was amazing. Almost like something out of "A Midsummer Night's Dream". It wouldn't have surprised me to suddenly see Puck or Oberon appear from behind the trees. A quite uncanny feeling permeated this solitary spot. It was completely silent with only the occasional bleat of a stray sheep. It was an experience neither of us will forget. Three hundred years had not changed it. The suffering and persecution seemed to emanate from the very rocks. The eyes that must have looked at that landscape were long dead. Only the landscape remained the same - frozen in time.

We made several more journeys up there subsequently. One quite early in the morning, with Conrad carrying Sarah on his shoulders and little Kate trudging along quite happily by our side. We later learned that, as part of their pilgrimage, the Covenanters carried each one of the very heavy stones, from the bottom of the hill to the very top, to set them in place to form the circle.

Every so often, Bill Aitken, our local Minister, held a sermon on that same spot and all the villagers would come and sit out in the open, to hear him preach. He was a wonderful man. Softly-spoken and very compassionate. He had time for everyone and was deeply loved within the community and with very good reason. He was a grey-haired, slightly-built man, with a quiet, unassuming air. His blue eyes twinkled behind his glasses but at first glance he appeared quite ordinary. Behind all this though, when you got to know him, his personality was quite the opposite. He had incredible wit, a razor sharp mind and once before his 'audience' in the pulpit, he gave an electrifying performance. He was a man of true, consuming passion and a desire to help and distribute love wherever he could. I think perhaps he was the most truly GOOD person I have ever met.

Conrad and he got on famously. Bill loved the theatre and they would spend many hours talking about it, the world, politics, philosophy, theology - everything. I think Bill fancied himself, just a little, as a 'theatrical' and it certainly came to the fore, when we all did a play together in the village hall. It was a Victorian Melodrama, called of all things "Hiss - The Villain", in which Conrad played dastardly Silas Snaker - a rascally banker (Boo), I played Sweet Lucy - a desolate daughter (Ahhh) and Bill played Bowler - the rascally banker's clerk - half baddy but ending up goody (Ooooh). He turned in a wonderfully funny performance and had the audience in stitches. We kept the village hall packed for three nights and had enormous fun every time. The audience participation was terrific as they booed and hissed in all the right places. It was a thundering success.

Bill's tiny church or 'kirk' as the locals called it, had plain white-washed walls inside, not ornately adorned. Simple and welcoming, it's ancient wooden pews smelt of beeswax and the flagstone floor was worn thin with the tread of centuries. It was a stout little place, solid and homely. Harvest Festival was wonderful. Wicker cornucopias overflowing with all manner of russet fruits and berries, marrows and pumpkins, turnips, tomatoes, flowers, bread and wine, all to be distributed to the needy of the parish. Nobody was forgotten.

Bill had baptised many, married many, seen their families grow - and buried many in the little churchyard at the side. He was totally concerned and interested in his 'flock' and found a true contentment in his work. His sermons were all invariably drawn from his own life experiences and didn't rely too heavily on Testament pushing. There were parties held at the Manse, that always seemed to be on the coldest night of the winter but were memorable for fun and laughter. It was open for Sunday School, The Women's Guild, The Old Folks' sessions and all manner of visitors. It was probably the only Manse that flew the Tricolour on the 14th July - a token of Bill's love affair with France. There was a good-sized garden attached and all the parishioners gave a hand tending it from time to time, alongside his gentle wife Moira. The produce was liberally shared in all kinds of generous ways.

Many years later, when Bill died, dear Moira wrote to tell us the sad news. She said - "It was so like Bill to go and die on a truly wild night, when the moon was full, with scudding clouds and a skein of greylag geese passing over". More than 300 friends and elders of the church turned out and carried him shoulder high to his final resting place.

"Whatever Bill Aitken's relationship was to his many friends, some things were common to them all - his transparent integrity, his absolute dependability, his irrepressible sense of fun and his gracious and cultured spirituality - and we give thanks for such a life - and such a friend". That was a moving tribute from Rev. Ian Robertson of Colvend, Southwick and Kirkbean and just about sums up the feeling we all had for Bill. He left a hole in all our hearts.

CHAPTER 14

Spring for us at Skeoch was a delight. The scurrying clouds alternately shading and lightening the first few flashes of the new green shoots. The sheep with their young, dotting the hillsides like miniature clouds come down to earth. You had the feeling you could blow them all away, with one quick puff. Winter snows and ice thawed and swelled the gurgling burns, throwing small stones and twigs along their shallow banks. It was a time of joyousness and after the grey and constant cold of the winter, a re-affirmation of re-birth and warmer times ahead.

Spring brought the swallows, curving through the sky like sickles, their ceaseless murmuring and chattering all day long. They would swoop low over the fields to catch the insects and buzz in and out of the cattle byres, making their intricate nests. The curlew, slightly mocking, would repeat its name all day long, while the skylark, ever busy, ever twittering on the wing, would plummet and bob over the burns and hills. Lazy, pollen-laden bees hummed around the garden and the adder, sinuously sly, basking in the heat between the crevices of the dry stone wall, would occasionally, languidly, slither into the garden. On one such occasion, Sarah, who was only two years old at the time, ran past bare foot. Our adder friend missed her by about five seconds. We felt this one was just too close to home and it was promptly dispatched by Conrad with the spade. The local farmers often complained that they had lost a lamb or sheep through adder bites and Jim told us that in the morning, by his back door, he would quite often see four or five of them, finishing off the contents of the dog's bowl.

Millie, our neighbour in the caravan, had not been well for some time. Her already frail body seemed to be shrinking by the week. She was permanently racked by a ghastly, rending cough, not helped by the perpetual roll-your-own stuck between her gums.

Our local doctor, was a kind and caring man. His surgery covered a very large area and he was obviously extremely busy. He always seemed to find time for everyone though and he dispensed a lot of sympathy and understanding as well as medicine. There was no appointment system at the surgery. You just came in and waited. Surgery was finished when the last patient had been seen.

The doctor had been coming to see Millie for some weeks, in her little caravan in the lane. Unlike her, she had finally taken to her bed. She complained to me of bad chest pains and a feeling of acute tiredness. Occasionally she would perk up when I came to visit her and her eyes would brighten if she spotted the odd jar of pickled beetroot I brought down for her. It was one of her passions and she would sit and eat it, just as it was, out of the jar - with a tot of whisky in the other hand!

The weather, along with Millie, suddenly took a turn for the worse. Constant rain and gales had battered us for some weeks. The dull thudding of rain on the caravan's metal roof, never letting up. The smoking fire inside gave precious little heat and Millie started to deteriorate rapidly. Finally she was moved into hospital and Paddy was left bereft and alone. Needless to say, without Millie there to control the reins to some degree, he drank more and more - to drown his sorrows. When his 'brew' money ran out, he would amble up to us for a glass or two of home-made bramble wine.

One evening in particular, just after we had had supper, there was a knock on the kitchen door.

"Have ye got a wee dram?" Paddy stood there, swaying, with tears streaming down his cheeks. We were obviously extremely concerned at his distress, fearing he had brought bad news of Millie. We told him we only had bramble wine but he assured us this would do fine.

Once we had him seated and he had a drink in his hand, all we could get out of him was - "Poor Benny. Poor, poor Benny. My Benny's gone. Aye - he's gone!"

Conrad and I looked at each other. Paddy was clearly .'three sheets to the wind' but we had no idea what he was talking about. Con decided to try and humour him.

"Oh dear Paddy. I'm so sorry. Your Benny's gone. How sad. Who is Benny, Paddy?"

The only Benny we knew of was Benny Hill and he seemed pretty fit and alive!

Paddy lolled on the chair and indicated that a drop more in the glass would refresh his memory. We obliged. He waited a suitable length of time, sure of our undivided attention, then commenced.

"My Benny, Conrad. I tell you, I loved that man. I did. I loved him. Now he's...gone!"

Gently Conrad said "Benny who, Paddy? This is the first you have ever spoken of him. Is he a relative?"

Paddy looked disdainfully at us both, swaying about precariously on the chair. His eyes glazed.

"It's my Benny Lynch," he uttered, swigging hastily. I looked at Conrad and whispered "Did he say Kenny Lynch? I thought he was a pop star".

Another half hour went by and the contents of that bottle and two thirds of another had disappeared. Paddy was still tearful but eyeing the room for signs of any more booze. Finally Conrad called a halt to the proceedings and helped Paddy firmly off the chair, telling him how sorry we were at his loss but that he should now go home and sleep it off. Unsteadily he got to the back door and started out the house.

"Do you think he'll be all right going down the lane?" I giggled at Conrad.

"He'll probably fall in the hedge a few times" he smiled. "Might sober him up a bit. But who on earth is Benny?"

Later we discovered, on good authority, that Benny Lynch was a boxer, who died some time in the 1930's!

Paddy took to visiting us more and more during Millie's stay in hospital, until the inevitable happened and we learned, sadly, that Millie had died.

Doreen, my friend and neighbour, obviously knew Paddy and Millie very well. Paddy had worked on Jim's farm at odd times of the year and they had lived in our lane in the caravan, for as long as she could remember.

Paddy had asked Doreen and me to attend Millie's funeral and of course we agreed, having extracted a firm promise from him the day before, that he would remain stone cold sober, at least until the service was over. He promised and crossed his heart, hope to die!... We were to take him by car to the cemetery in Dumfries and had agreed to call on him at 11.00 a.m. on the morning of the funeral. We were, therefore, very apprehensive when we knocked on the door of the caravan that morning. The cursing from within did not bode well and when Paddy finally emerged, we knew why. He had managed to put on his good suit and had half shaved, cutting himself on the face in the process, a few times. He had stuck small pieces of cigarette paper in the nicks in an attempt to stop them bleeding. His tie was flapping around his neck and his collar was bloodstained. His cap was askew, he was glassy eyed and ready to take on Floyd Patterson! So much for being sober. We straightened him up as best we could and propelled him to the car. All the way into Dumfries, he cursed and swore, threatened and cried. We alternated between cajoling, imploring and threatening him ourselves - all to no avail. Somehow we managed to get through the service with Doreen and me taking it in turns to clap a hand over his mouth before he uttered oaths, our voices getting louder and louder as the hymns progressed. The finale came when we were at last outside. Paddy was to have been a pall bearer but the fact that he was a good two feet shorter than the other three men, plus his obvious state of intoxication, it was decided it would be better if he were replaced. This did little to improve his state of mind and he emphatically and abusively insisted that he help lower the coffin into the grave. Reluctantly he was allowed to do this, the Priest standing by, watching with some alarm as, instead of lowering the rope gradually, along with the others, Paddy just extended his arm into the void. Just as the Priest was about to sprinkle holy water...it was a deep hole and Paddy was a short man and more the worse for wear. He teetered on the edge, swaying. His cap, which up to now had remained slightly askew but firmly placed, now gracefully slid on top of the coffin and Paddy was fast disappearing after it. By quick thinking Doreen and I grabbed his pants in the nick of time, just before he, too, disappeared. He let out a string of profanities and a mighty torrent of four letter words, successfully drowning out the poor priest, who by this time was ashen faced. We pulled him up between us, until his feet left the ground, put a hand over his mouth and stood there with him balanced between us, feet dangling, while his cap was returned. Finally, Millie was safely sent on her way. On what should have been a sober occasion, in retrospect, I am afraid I probably shed more tears of mirth than sorrow. Millie's ills were past and we were relieved that at last she was free from a very wretched life and free from pain. I couldn't help feeling, as we walked away from the leafy cemetery, that she was probably still with us, looking on and admonishing Paddy with ..."Yon bleeding auld sod's still at it, couldna' stay sober even on me funeral!" and one of her lovely toothless grins.

Several weeks drifted by and after an initial 5 days booze-ridden rampage, Paddy took up residence once again in the caravan. After a while though he got very despondent and dejected. He seemed totally unable to come to terms with Millie's death and talked more and more of moving on. He took to coming up to see us more frequently, for a glass or two and a look around. On one of these jaunts, he told us how lovely he thought our donkey filly "Rosie" was and ultimately asked if he could buy her off us for company. We deliberated for a while but in the end thought it might help him in his loss and finally agreed. All appeared to go well for several weeks, until one night he came up to Skeoch and said that after much thought he was moving on. Sitting in our little kitchen with a large glass of something intoxicating, he said it was an old gypsy custom, that if a partner died, then the caravan and all their possessions were burned. In Millie's case, he had in fact burnt all her few pathetic clothes and bits and pieces but he did not want to burn the caravan. He asked if we wanted to buy it and at length we agreed, saying that as soon as we could get the tractor down to move it, we would tow it back up to the farm.

The caravan itself was very pretty, with tiny stained glass windows and a stove pipe chimney. It needed a good paint and clean up but we thought it might convert into extra bedrooms maybe for the children in the summer, to sleep out. They were very excited at the prospect of that. I felt sure that Millie would approve of the whole idea.

Paddy finally came up to say goodbye, never telling us where he was heading. I think perhaps that he didn't have a clear idea himself at that time. There were tears and much handshaking as we wished him all the best. He had sold Rosie the donkey already...and King the dog. He was travelling light.

One afternoon some weeks later, we finally wandered down the lane with the tractor and hitched the caravan on to the rear tow bar. Conrad gently eased the tractor forward. The caravan refused to budge. He tried again and again. We spent hours levering, pushing and coaxing, trying to get it to move an inch. It appeared that the years of 'settling' had made it absolutely impossible to move. We tried digging the wheels out - no luck. Finally in desperation, we had to admit defeat and left it where it was, in the lane.

We didn't hear from Paddy again and life resumed it's

implacable course. With so much to do, we were kept constantly busy with the herd and all the other tasks that queued up every day. Then one afternoon, some months later, Doreen came racing up the lane, all hot and bothered.

"The caravan," she blurted, "come quick. It's on fire". We sprinted back down the lane, in time to see clouds of acrid smoke billowing up and an ominous crackle. By the time we finally reached the spot, we saw the whole thing go up in a mighty explosive blaze. The fire completely engulfed the tiny caravan. Flames licked through the shattered windows and open door. There was absolutely nothing we could do. By the time the Fire Brigade arrived from Dumfries, the van would be well and truly gutted anyway. We just stood back and watched mesmerised as bright flames and black smoke billowed skywards, singeing the branches of the overhanging trees.

How the fire started we never knew, although we had our thoughts on the matter. Nothing whatsoever remained of the van except the hub caps where the wheels had been. So, Millie had her send-off gypsy fashion after all. I can't say we were really sorry. It all seemed somehow appropriate and a fitting end. We were left, after all, with a wealth of memories from a very special person.

Within six months, nature had worked her usual miracle and covered the entire spot with weeds and plants, including a fine bed of rhubarb which we had never noticed before. All traces of a previous life covered over - except for a few jewel-like shards of stained glass window, occasionally catching a glint of the sun and twinkling through the grass.

Some time later the police came up asking the whereabouts of Paddy in connection with the burning of the caravan. We had no idea where he had gone and we believed that in his present state, it would not be too long before he joined Millie. We never saw or heard from him again.

CHAPTER 15

There is something wonderfully 'heady' in purchasing your own car for the very first time, albeit an old banger. Annabel was a case in point. Conrad had been offered a very lucrative part down South again and it meant that he would be away for a period of about three months all told. As he needed our car and I can't ride a bike, it meant that I would be severely restricted while he was away. There was no local transport system to get into Dumfries unless I walked the two miles to the road end, which in itself would be OK but with two small children and then bags of shopping, not something I wanted to undertake. There was also the fact that, with our increasing animal life on the farm, various bags of feed had to be got in, so it was clear that some alternative form of transport would have to be found.

Some time previously, I had been introduced to a nice young man, Davey, who owned a rather racy car show room in Dumfries. He was a Mister Fixit (another David!) of the transport world and I promptly hot-footed in to see him. Conrad suggested that I hire a car for the three month duration, while he was away.

Davey grinned when I told him why I had come in.

"Have you not thought of buying a wee car, Jennie?" he queried. "I have a little beauty here".

It would be too expensive, I told him. We did not, at the moment, have a budget that would support buying a car, particularly when we already had one.

"You will do when you hear what she's going for," he said.

I followed him to the corner of the showroom, somewhat intrigued. Sitting there, in all her maroon glory, was an extremely ancient Austin A30, complete with semaphore indicators.

"Thirty pounds and she's all yours," Davey smiled. "It would cost you more to hire one for three months, Jennie".

"Does she go?" was the next question.

"She's more durable than I am," he quipped. "She'll take you from one end of the country to the other".

I was, of course, completely hooked by this time. A few phone calls, Tax and Insurance taken care of in what seemed a matter of minutes and, somewhat dazedly, a half an hour later found me behind the wheel and driving out into the Dumfries traffic. The elation I felt driving my little car along the beautiful summery country lanes back to Skeoch, was something unequalled. But I was slightly perturbed when I wound down the window to let some air in, to find that the window, with a gentle thud, disappeared completely inside the door and no amount of coaxing seemed to wind it back up. No matter, it was a small price to pay and it was indeed true, she had a most sturdy little motor. Dreamily I pulled into a petrol station and flushed with my new success, filled her up. As I drove up the lane towards Skeoch, I was just a little alarmed to find that my fuel gauge only registered a quarter full. Drat, I thought, obviously the gauge doesn't register properly. Once inside I rang Davey and told him.

"Oh that's a wee thing I forgot to tell you, Jennie" he laughed. "Don't put more than two gallons a time in her, because she has a wee hole in the petrol tank, just about the two gallon mark".

Oh well, what could one expect for £30 . It had four wheels and it went!

The children were delighted with my new purchase and she in fact ran beautifully. I never had a moment's problem with the engine, except the odd fan belt and I do have to confess tha after a few months, the door catch on the driver's side finally gave up. This was all right if I was driving in a straight line but the Scottish lanes are pretty notorious for their somewhat hairpin bends and the only thing I could do then was to tie a length of bailer twine to the inside of the door handle and tie the other end to the steering column. This saved me having some very near misses, going round corners with oncoming traffic, when the door flew open. I would hasten to add that this was all pre MOT. Annabel, as she became known, served me wonderfully well. She would pootle along at 40 mph all the way to pottery classes on Tuesday nights, some twenty miles away, which I had taken up in great anticipation. The freedom of having her was wonderful and as long as the bailer twine didn't snap, I was all right. In fact I used her so much, that various parts of her body eventually dropped off and it was neither safe nor comfortable to drive her. Even so her days were not numbered. She was most useful to have in the Autumn, when I would drive her across our fields down to the woodland and stack her up with logs which Conrad had cut up and drive her back up to the house.

So many busy springs and summers and the autumn was no exception. We would gather baskets of blackberries or brambles as they were called, which grew in thick masses along the lane and in our woods. The ripest and best of course, were always in the middle of the bush. Later we would wander down in the early evenings, to shake the nut trees and dawdle back with cold feet and usually a fair crop of sweet cob nuts for Christmas. Bramble wine was one that we first tried our hand at. We used to store all the home made wine upstairs in the attic bedrooms (when not in use). It was always quite warm up there and the wine could bubble and pop away to its heart's content. We tried various sorts. Dandelion (of course), barley, apple. In fact if anything could be made into wine, we tried it. One night, desperate for something stronger than coffee, after a particularly gruelling day injecting cattle, we decided to see how the bramble wine was coming on. We sampled some that had been made about four months back (hardly mature!) but in desperation .... it was absolutely foul. I decided it would have to go and I even had thoughts about what it would do to the septic tank! As ever, I seemed so busy, weeks went by and I never got round to disposing of it. About a year later, having a major clear out in the attic, I came across two dusty demi-johns and brought them downstairs to finally tip them away. Before doing so we decided to have one last try, not holding out much hope but also noting that the contents seemed beautifully clear. It was delicious. In fact one of the best wines we ever made and obviously one that just improved with age, being mellow and fruity. Unfortunately the two demi-johns were polished off all too quickly, with nothing nearly as good to replace   them   with.

We had a load of sloes too, which were torture to gather but worth the effort when the first drops of sloe gin warmed you up after a cold day outside, but one always had to BUY the gin in the first place!

Our doctor once said "When you stop giving your cattle pet names Jennie, you will be a proper farmer". He was of course totally right but my life, coming from London, actors, theatre etc., was not a very good basis on which to model oneself as a farmer. Nothing could be more dissimilar. I bought my eggs, meat and veg from shops, like everyone else. I was not then unduly concerned about how it came to be in the packet, or what went into producing it. Calves were sweet, something you smiled at when you saw them in a field in the country! When we first started out, therefore, one calf in particular, found her way into my heart like no other.

With our second consignment of two week old heifer calves, among them was a very pretty dove-grey and white Charollais calf. She was more like a baby deer than a calf and she took to me immediately, nudging my hand with her little wet nose. We were delighted with her and I called her Feiline (would you believe from the film Bambi!!) This shows you how ill-prepared and naive I was, to be a farmer.

Feiline was lively and alert but had one fault. She flatly refused to feed from the bucket. She was one of our 'problem' children that needed to be fed from the bottle. This in itself, was no big deal but when after the first week I went to feed her as usual and she would not stand up, I started to worry. I tried to get her to her legs but she could not stand, just sinking slowly to the ground when I let her go. She seemed absolutely fine apart from this. Showing no signs of fever and still taking the whole of her bottle with her usual alacrity. The next few days were the same, no loss of appetite but she could not stand. I rang the vet and got Mr Murchie. He arrived up and examined her. He was very gentle and very sympathetic. He gave her some injections and arranged to call up and see her a few days later. He wasn't sure what was wrong. Still she showed no signs of improvement but every time I went into her pen, she showed great signs of excitement to see me and drank all her milk.

I spent hours with her, massaging the muscles on her legs. I put a sack under her tummy and hauled her to her legs in an attempt to try and help her to stand but the moment I let go, she would collapse to the floor.

Mr Murchie called again. We changed the treatment. He was encouraging and kind. She stayed like this, all in all for about 6 weeks, during which time she seemed quite content just to lay in the straw and drink, not appearing to be in any pain or distress but she was absorbing more and more of my time. I prayed for a miracle, that I would go in one morning and find her standing. It was not to be. Towards the end of the 6th week, Conrad had to go away for more work. We were in no position to turn work away and it was a tough schedule and work load for him. The farm was costing us every penny we had, with everything going out and nothing coming back in.

My mother and sister arranged to come over and stay with me while Con was away, to give me a hand with the children.

One afternoon I went out to see Feiline. She was suddenly not so good and as I moved her to put fresh bedding under her, I discovered a sore under her chest. It was obviously causing her a lot of discomfort. In tears I rang the vet. The time Mr Allison came up. With his usual brisk manner, he looked her over. There was no beating about the bush.

"I'm sorry, she has a form of paralysis, that will never get better" he said. "She could stay like that until she's twenty but she will NEVER stand again. There's absolutely nothing left that I can do - except put her down. That would be the kindest thing". I was distraught. She looked so healthy. She gently nuzzled my hand.

"So," Mr Allison said, "let's do it now. Just sit there with her and make a fuss of her. I will give her an injection and that will be it".

I sat on the straw with her little grey and white head in my arms, tears streaming down my face, while he administered the injection. She didn't struggle. It was done in a minute.

He looked at me kindly. "Do you want me to arrange for the carcass man to come on up and collect her"? he asked.

I shook my head and thanked him. "I'll arrange that myself" I said, trying to pull myself together.

I saw him to his car and went in doors. I could not bear to have her unceremoniously hauled up on to the carcass wagon. It was a depressing and ignoble end to a valiant little creature. The meat wagon, as it was locally known, was horrible. We had, in the past, had to resort to using it when we had a fatality with a heifer, fortunately not often but it nearly always seemed to arrive up with a full load. It would lumber up the track, a huge open sided lorry, displaying various dead carcasses. Full grown cows and calves in various grotesque poses, tongues lolling out, eyes glazed. Some bloated, some badly injured. It was a hateful load. "Bring out your dead"!. The driver would slide an iron winch over the dead animal's hooves, press a lever on the lorry and slowly the carcass would slide across the concrete and up on to the lorry. It hung, suspended for a moment until he had it in position then, with another press of the lever, it would drop on top of the others. I hated it.

Foolishly and sentimentally though, I decided to bury Feiline - down in the woods. For this, I placed her in an old cot blanket and helped by my sister, put her in the wheel barrow, which took some time I can tell you because an eight week old calf is no light weight. I finally managed to wheel her into the yard and started to descend down the hill, through the bottom field, towards the woods. With the weight of the barrow and the slope of the field, my momentum increased. I was gathering speed at an alarming rate. It was as much as I could do to stop the whole thing running away with me. Red-faced and panting, I finally reached the wood and found a spot I thought would be perfect. Swinging the pick axe into the earth a few moments later, I was met with a dull thud. Scotland is built on rock! It finally took my sister Mary and myself about three hours back breaking graft to get a hole deep enough to bury Feiline. We ended up piling a mountain of rocks on top for good measure, said a tearful prayer and wearily went back to the house.

The death of Feiline taught me a valuable lesson. Although I never lost my softness or compassion, I never again allowed myself to become too emotionally involved with the calves.

With hindsight I should, of course, had her put down in the first couple of weeks and probably saved her and myself a lot of pain. Who was it that said "Through pain we grow?" It is difficult to deliberately change your attitude towards things, especially emotional ones. Some of us always seem to choose the hard way; I was not a natural born farmer. Doreen, I know, would have been much more practical.

CHAPTER 16

We had known for a long time that the huge corrugated iron roof on the large barn that acted as a garage was unstable. There was a lot of corrugated iron at Skeoch, much of it none too well fixed. On a reasonably windy day, a corner of the garage roof would lift a little, clatter and flop down in a desultory fashion. It was one of the many jobs we would 'get round to' when we had a moment to spare. Needless to say, the time was never found and we just got used to the odd clang, reminding us that it needed fixing. However, the wind finally forced our hand one morning at about 6.00 a.m. A gale had been raging around for several days and what had started as just a corner clanging away, became several sheets lifting. By now the wind was so strong that we couldn't get out and get on the roof to start any repairs. After this particular night, neither Conrad or I could sleep, so we got up at about 5.30 a.m. to have a cup of tea. Peering into the wind-tossed yard, at first morning light, we saw the garage roof gradually start to lift. Sheet upon sheet of corrugated iron, once bolted together, rose in slow motion and started to spin, like a giant plate, towards the kitchen where Con and I were rooted to the spot, staring in terrible fascination. With a tremendous crash, it sliced through the air and straight into our electricity supply, cutting it off immediately. All the lights, heating - and worse - the deep freezer. We depended on it entirely for all our needs. Without it the pump in the field would no longer take water to the reservoir. We cooked by it, heated our water with it. It was our life line. So it was imperative that the supply be connected as quickly as possible. We turned off the mains immediately and had to wait for the wind to abate before Conrad had the unenviable task of battling out in the still strong winds a little later, to clear a path through the debris. With enormous ingenuity, insulating tape and various pieces of cable, he succeeded in re-connecting the circuit but only after a huge amount of labour, swearing and effort. Not for the first time did I thank my lucky stars that he was a very practical man as well as an artistic one.

It was amazing that the strength of the wind had taken the roof like a piece of broken biscuit and tossed it carelessly away. Another two feet and it would have hit the kitchen window and probably killed us. The Western gales were very treacherous.

We then had the unenviable task of repairing and replacing the actual garage roof. Closer inspection the following day, revealed that many sheets of iron had withstood the worst of the wind. Some were just hanging by a thread but a lot had to be replaced. The pieces that were torn off were totally irreparable, having wrapped themselves around poles and objects on their journey to the kitchen.

We went to Dumfries and bought a supply of five inch long bolts with collars and nuts on the end and new sheets of corrugated iron. With just the two of us to do the repairs, it was to be a hard task. I would clamber up on to the garage roof by step ladder and with Conrad's help, place a sheet of iron in the right position. He would be up a ladder in the garage below. I would push a bolt through a hole in the iron that he had previously drilled and then he would fix the collar and nut from underneath, to the girders. And so we went on. It was an arduous task and one that, as ever, took far longer than we anticipated and was completed, in the end, after about a year. At that stage we seemed to be constantly running from pillar to post. No sooner repairing one thing, than some other disaster struck and we would drop what we were doing to try and mend something else, often totally exhausted and in not such a good humour!

It was mid-April and Con and Kate had just celebrated their birthdays. Kate's birthday being just one day after Conrad's. Even so, the heavy coating of thick white frost on all the trees and shrubs, showed no sign of lessening. The pond was still frozen solid and a biting Easterly wind almost took your face off the moment you put a foot outside the door. It was raw to the extreme but, as was so often the case, incredibly beautiful. The landscape hard and frozen was held in the Ice Maiden's grip. Sounds became muffled and the air was so cold that it hurt to breathe but the skies by day were a brilliant blue. Winter had started in the September previously. Eight months of cold, snow, sleet and frost. We were now longing for the summer and desperate to see a bud, a thaw - something.

Doreen came up for a coffee one morning. She adored Kate and Sarah and I am glad to say, spoiled them, just a little! She said that her sheep dog Pat had just produced another litter of puppies. The conversation took a turn towards the fact that we hadn't, as yet, got a dog. It certainly made sense to eventually get one. We had plenty of space and when Conrad was away, it would make me feel more secure.

We decided to come down and look at the litter but

decided also not to tell the children for the time being but surprise them.

Dear old Pat was laying on a pile of sacks, in one of Jim's outside cattle sheds. She wagged her tail limply and looked apologetically at us. She was a wonderful sheep dog. The traditional black and white border collie. Every year she worked with Doreen at lambing time and then, throughout the year with Jim, herding sheep, cattle, running tirelessly day after day. In between times, she had litter after litter of puppies; some of which were found homes and some of which were drowned. We bent down to give her a stroke and look at her offspring. Three tiny black and white puppies were nuzzled into her groin, their eyes still closed. Faced with a decision to choose one, I was in despair. They were all so adorable and the knowledge that whichever one I chose, the other two would probably be drowned, made the choice almost impossible. It was not that Jim was an unkindly man, far from it but hill farming in Scotland is a very hard life. There is not much time for sentimentality on animals. They were just functional. A dog was kept to work not kept as a pet. Cats controlled the rodent population and all were treated on the same level, whether it be a cow, a bull, hen or dog.

Finally we plumped for the one dog in the litter. He was a miniature version of his mother and by the looks of him, already very greedy. We said we would come and collect him in a few weeks time when he was weaned. We called him Sam. The children knew nothing of this - Sam was to be a surprise. They adored him from the moment they set eyes on him and for some time he made a wonderful companion, getting into all manner of puppyish mischief. By the time he was a year old, he was ready to go outside. He was not an overly-demonstrative dog and preferred to be outside rather than inside so eventually we housed him in the large store shed by the side of the house. He had a long lead and could come in and out as he pleased. In the summer he always preferred to sleep on his cushions outside the shed, where he made an admirable guard dog. He was totally obedient to Conrad and totally disobedient with me!

Now bread-making was never one of the things I mastered in all the years before and during Skeoch. Something I have now remedied, I am glad to say, but any attempts previously invariably turned out leaden. I remember well at one time, being so intent upon producing 'wholesome' natural bread, that I pushed on every unlucky member of my unwitting household, piles of barley buns. These scrumptious-smelling morsels looked as if you could build the next lot of Barratt Homes out of them and they weighed as much as well but having our own 'bruiser' (a machine that crushes whole grain) made me slightly fanatical about trying to produce them. We really did want to be as self sufficient as possible and this was my weak link! The actual machine (which weighed about a ton) was basically quite easy to operate. A series of various pulleys and belts rumbled to life once the electric motor was turned on. When it was going, you tipped your whole grain into the funnel at the top, a huge stone ground the barley into flakes, husks and all, and it was collected into a waiting bucket below. It was a laborious task and also kicked up such a din that I was deaf for about ten minutes afterwards, but the cost of buying in already ground barley for the winter feed was astronomical. The chickens enjoyed the barley buns most I think. They were lethal for human consumption but would have provided superb training for a discus-throwing family. One of my barley buns, aimed at ten paces, could have felled a sixteen stone man!

All in all we ate very well. We had a constant supply of

fresh lamb, pork and beef in the freezer, along with all our own peas, broccoli and vegetables. My chickens usually died of old age, simply because on the one occasion that Conrad tried to wring the neck of a rather frail chicken (which we thought could at least be used as a boiling fowl), after hanging it up by it's feet and duly ringing it's neck, it promptly appeared to come to life again twenty minutes later and had to be 'finished off' with the .22!

In the summer months, I occasionally did a little bed and breakfast, which all the local farms did if they could, to help supplement their rather meagre income. This enabled me to have a Cash and Carry card, which was a boon, particularly as we lived so far away from shops. Regularly we would go for a 'spend'. Things like loo rolls, boxes of baked beans, whole Scottish cheddar cheeses and drums of cooking oil, were bought and loaded into the van. Over the course of a year, it provided us with quite a saving, as well as the convenience of not running out of things. When you live in comparative isolation, the siege mentality tends to take you over.

I was registered with the Scottish Tourist Board and also 'Where to stay in Scotland' which sounds rather grand and although our accommodation was sparse, we never had a shortage of bookings in the summer months. In fact, on one occasion, just the opposite.

I knew that I had an adult couple arriving some time during this particular Saturday. The attic bedrooms were prepared and I was looking forward to their arrival. What I was not prepared for was the phone ringing at about 11.00 a.m. and a friendly voice saying ..."Oh, hello Mrs Phillips. Just thought we would give you a tinkle and let you know we will be arriving at about 4.00 o'clock".

I smiled. "Ah, Mrs Kennedy, how nice of you to let me

know. All is ready for you".

There was silence at the other end for a moment or two, then "Hello - are you there? - er no, it's not the Kennedy's dear - it's the Bartlett's. You know, Mr and Mrs Bartlett with Simon and Jessica".

I felt the sweat break out on my neck. "Ah yes, Mrs Bartlett of course" I lied. "Just the four of you isn't it?"

"Yes dear, that's right. You got our cheque all right didn't you?"

Then it all came flooding back. Yes I had received the cheque, yes I had confirmed the booking - it was the Kennedy's I had completely forgotten to cancel. It was a scene from Fawlty Towers for real.

So here I was - double booked. Four adults and two children, with only one double bed, Conrad's and mine, two single attic bedrooms and our children's twin bedroom with bunk beds. What to do, what to do?

Frantically I rang Cathy, my girl friend, who had two children of her own, and explained the situation. "No problem," she said. "I'll take the girls for the night". Wonderful. Two down - two to go!

I zipped around our bedroom changing sheets and duvet covers, grabbing underwear out of chests of drawers and dumping them in the spare box room, which opened off our bedroom. This room also, thank goodness, had two large, rather moth-eaten floor cushions, which had seen better days in the swinging sixties. It was, by all accounts, the junk room, housing everything from fermentation locks to old 'might come in useful' lampshades, tennis racquets, plimsolls, ironing boards, chairs that needed cane seating repaired etc. Fortunately though, this room did have its own entrance, through French doors, into the garden. We threw a spare duvet on the floor cushions and locked the door between it and our bedroom.

All our guests arrived in due course and had their evening meal together in the kitchen. Fortunately they all seemed to get along wonderfully with each other and finally they retired to our sitting room to watch television. By this time Conrad and I were knackered and starving. Obviously I couldn't walk through our sitting room from the kitchen then through our bedroom to the spare room, with trays of food, so I left through the kitchen door to the garden, balancing our supper on a tray and went round the side of the house to the back garden, to enter through the French doors. Actually it took on the proportions of a French farce. Imagine my embarrassment when mid-trek, I encountered one of our male guests happily 'watering the roses' by the light of the moon. Having spotted each other it was too late for me to retreat - and too late for him as well. I mumbled something about it being a lovely evening, carefully averted my eyes and bolted in through the French doors.

Conrad and I sat eating, amid the junk, with a tumbler of dandelion wine to ease the situation, thanking God that we only had to do this for the one night. Full and sleepy, we squeezed ourselves onto the rather mangy floor cushions. We spent an awfully uncomfortable night but at least we did it and the next day, only one person was any the wiser!

We celebrated our extra income by buying some sacks of molasses for the cattle, a new garden fork and some giant balls of twine for the bailer. Heady stuff!

Being a bit more mechanised, we decided it would be in our best interests to grow our own barley and cut down on bought-in animal feed stuffs. The coming winter saw Conrad proceed to plough up an old grass field, that led gently down to the wood. It was a four-acre field and also Conrad's first attempt at ploughing. He suffered 'Ploughman's Neck', as it was jokingly called, for days afterwards, brought about by constantly looking over his shoulder at the furrows to see if they were straight. His back was a mass of Cleg bites that had bitten through his shirt. He said he felt really a farmer when, looking behind his tractor, he saw appear from nowhere a flock of sea gulls hovering over the freshly turned furrows which finally, just as mysteriously, disappeared. Having done all the necessary preparation of the field in the winter, it only remained then, by the spring, for us to actually plant the seed barley which we had ordered.

DK came up with the answer, in the form of an ancient 'dibbler'. This ingenious piece of equipment was a large, long trough, which carried the seed barley. It had funnels evenly spaced apart, leading from the trough to the ground. There was a narrow platform attached to it, on which one stood and two large levers, one at each end. This whole contraption was hitched to the tractor and we were off - Barley Boadicea style! Conrad driving the tractor in front and myself, precariously perched on the narrow platform, as if on a chariot. I had to push the huge levers down to release the seed barley, once we were in line, then jump off and pull the levers up every time we reached the end, to stop the barley pouring out. Once back in line, push the levers down and so on. As it was old machinery, the trough seemed to run out of barley almost at the end of every two rows, making it a very slow task. After a few hours of this, my arms felt as if they were made of lead and with so much jolting about on the platform, my body felt as if it had been dropped from a height without a parachute. Conrad had no problems with this - all he had to do was drive the tractor. The whole process lasted for a few days until we finished, at the end of which I was absolutely worn out. It was just as well that we planted the barley when we did, as a most foul spate of weather arrived, locking us into the house for days on end. We squelched up the muddy lane, only to feed the sodden, freezing cows and throw corn to the bedraggled chickens. The skies were very heavy and grey and refused to budge. It dumped over Skeoch for days, waterlogging some of the lower pastures and making us all thoroughly bad-tempered and fed up. The children, unable to go out and play, were hyperactive inside the house and my patience was beginning to wear thin. Conrad was convinced the seed barley would rot in the ground if we didn't have a break in the weather soon, so that was an added worry, along with a cow that had contracted pneumonia and required almost constant monitoring and medication in one of the byres. Just to add to all this, I went down with a bad dose of flu and was finally, reluctantly, forced to stay in bed by the doctor, with a rampaging temperature, vomiting and gripey stomach. Conrad had the whole thing on his back, farm, children, sick wife, sicker cow and non-stop rain! After what seemed like an eternity however, the weather finally broke, to be followed by a spell of balmy breezes and the long-awaited sun. What a relief and, as if to prove that every hundred grey clouds have a silver lining, a faint green haze started to show on the barley field.

Pheasant life grew in abundance at Skeoch, primarily because a neighbouring estate used to rear the young chicks by hand, ready for the gaming season. I always used to think it such a heartless pastime, that the young chicks would be hatched (usually by bantam hens because they make better mothers!), kept in compounds and hand fed daily until of age to be driven out of the compounds by the beaters and on to the guns. The birds had grown accustomed to man feeding them every day and become quite tame. So when the gaming season started, I must confess, quite a few of these birds found sanctuary in our neck of the woods. We rarely ever ate any and those we did, were usually given to us by a neighbour. On one occasion a badly damaged, gloriously coloured large cock pheasant fluttered on to our fields. He was so badly hurt that, out of kindness, Conrad had to 'finish him off'. Ever practical, Con suggested that we hang it for a few days in one of the cattle byres and we could either deep freeze it, or use it as we pleased. Accordingly we trussed him up, some good eight feet from the ground, suspended with a stout piece of bailer twine hanging from a nail in the wall. After three to four days, we decided to have the pheasant for dinner and I went out to the barn to fetch it in for plucking. All that remained were a bunch of feathers, feet and head, scattered over the barn floor. A very bloated, satisfied-looking Zebedee came out of hiding some three days later! I am just amazed at how he managed to run up the wall and pull it down - but he was a big cat!

The only other time that I have seen Conrad take a pot shot at a pheasant was with the appearance of the barley. From our bedroom window we could see down to the wood and consequently the field in front, where the young barley shoots were just starting to appear. Strolling through this, one fine morning, was the largest, sleekest cock pheasant I have ever seen, resplendent in a giddy array of bright feathers. He sauntered along, pecking at the tender shoots and gobbling them up as he went. Conrad was understandably outraged. Opening the bedroom window, he yelled his head off at the offending bird, who in turn gave him a very haughty gaze back...and carried on eating. Con aimed a missile, which missed. Still the bird ate. Finally, in desperation, the .22 came out. By this time the children had marched into the bedroom, having heard their father yelling blue murder.

"Dad, you can't shoot that bird," a disgruntled Kate shouted.

"Try me," was the cryptic reply. He loaded up, aimed and fired. A small puff of dirt kicked up by the pheasant's foot. He stopped eating for a moment, stared interestedly at the dust and carried on eating. Again Conrad aimed. The children started dancing around to distract him. Another pocket of dust puffed up by his beak. The pheasant looked up, stared at the sky, stared at the ground, stared at Conrad and sauntered off, amid then, a volley of bullets, stopping occasionally to pick at a barley shoot. In no way was he at all ruffled, or even marginally alarmed. He was like the man in the ice cream suit! Undoubtedly the coolest pheasant ever. As for me, I was doubled up in the corner of the bedroom in hysterics, along with the children. Conrad turned round to look at us, a very stern expression on his face.

"I do not, for one moment, think it at all funny to see a pheasant eating our grain". The children and I sobered. "This is our future livelihood here. If we can't make it work, we can't stay". He spoke with mock severity and of course what he said was true.

Kate gazed up at him. "Dad, I thought you were a gunner in the Navy in the war," she said, edging towards the door.

"So I was," he said smiling. Kate whispered something to Sarah and prodded her. "You're a lousy shot," Sarah giggled, and the two of them fled out of the room in high glee.

Conrad came over and put his arms around me, grinning.

"I wasn't aiming to kill it anyway," he said, giving me a hug. "Just frighten it a bit".

"Well, you succeeded there, darling," I smiled, "It's in the garden now - eating the peas".

CHAPTER 17

Orlando had turned into a fine young colt and trying to get a head collar on him at the age of four months saw me hanging on for dear life as he dragged me on my stomach up the field - determined not to have the head collar on. He was extremely strong. His grown-up coat was a bright burnished copper, with a little white flash down his face and three white socks. He was beautifully proportioned and my pride and joy.

By the time he was a year old, the wife of our other neighbour, suggested that we have him castrated. I didn't really want to do this at that time. He looked as if he had the makings of a fine stallion. However, she said that she had a load of young pedigree Welsh Mountain mares in a field that adjoined ours and she was worried in case 'anything happened'. She seemed concerned that Orlando might sail over the dyke one day and sire the lot of them! Reluctantly we agreed to have him 'done'. The vet came up one morning and injected poor Olly with a lethal thing called 'Immobelin' which acted immediately. Orlando just dropped, as if pole axed, while the vet proceeded to castrate him. The vet kindly handed the testicles to Conrad. Neither of us were entirely sure what we were supposed to do with them...so Con took his knife from his back pocket, dug a small hole in the earth...and ceremoniously buried them in the field.

After he was stitched up, the vet told us to hold Orlando down for as long as we could while he injected him again with an antidote called 'Revivin'. Olly struggled and we held on to him for as long as possible. Finally we let him go and he struggled somewhat shakily to his feet. We felt so sorry for him. Poor little chap. We slowly led him to the tiny paddock by the side of the house, so we could keep an eye on him. It was a pretty little place, full of long grass. By the afternoon he had worn a path through and flattened it, going round and round, in a very painful, slow plod. Con winced every time he looked at him! So, the deed was done and a few days down the line saw him almost back to his old self. Pity though, he would have made a fine stallion.

At one of our monthly visits to the cattle market in Dumfries, while strolling round we saw, alone in a pen and complaining very loudly, the most beautiful little grey and white Charollais calf we had ever seen. At two weeks old she was already strong and healthy and obviously totally distressed at being parted from her mother. As we approached her pen she came over and I put out my hand to stroke her. She immediately started furiously sucking my fingers with an ecstatic look on her face. She was absolutely starving. By now I was hooked. I really couldn't bear the thought of her going to anyone else and although we hadn't come to market to buy anything, Conrad took one look at my face and..."I suppose we could try a bid for her," he grinned. "She would just about fit into the back of the van".

Although we were now rearing only bull calves, since the disastrous outbreak of Brucellosis, our pasture had subsequently been rigorously tested by the Ministry of Agriculture and pronounced "clean' and Brucellosis free. We thought that perhaps we could have another go at breeding the odd calf again. This little calf could make a fine mother. She had come from an accredited herd, which meant she was free from the disease as well. It was worth another try.

When she was ushered into the ring, we were first in line with the bidding. Far too eager, and I suspect we probably paid over the odds for her. We couldn't wait for the sale to end and get her into the back of the van, which was accomplished fairly quickly. She continued to complain loudly all the way home.

Juno, as she was promptly christened, presented no problems with her eating and no problems with her health at all. By the time she was twelve weeks old and on solid food, we were immensely proud of her and when she was finally turned out to new grass, her exuberance and antics were a joy to watch. She had become very special to us and had sailed through all her stages of growth perfectly. A model baby and a textbook case.

By the time she was eighteen months old, she was ready for 'artificial insemination' and we hoped she would be a model mother and produce a fine calf herself. One morning we called up the A.I man from the Ministry of Agriculture. Juno was in season and we penned her in, ready for his visit. When he arrived we herded her into the crush. She was quite amenable to this and strolled in with no hassle. He bustled about, donned his rubber gloves and came over, carrying a long tube with a plunger at the end. This contained the sperm of a good pedigree bull and very expensive it was too! He inserted his gloved arm into her with the precious phial and within two minutes the whole procedure was done. Juno didn't seem to mind at all and just gazed at us, through long eyelashes, as if it were an everyday occurrence. The A.I. man left and we returned Juno to the field. We were mightily excited at the prospect of what should be a grand calf and waited for the signs. The only sign that we had was that Juno was in season again some time later. Obviously this meant that the prized calf was not going to be this time around. So again we rang the A.I. man and the whole performance was repeated. We had about four or five attempts in the end, to get her into calf, all to no avail. The vet came up and examined her and could find nothing wrong. She was in prime condition. There seemed no reason why she never became pregnant. It had been successful with all our other cows - sod's law that they were the ones we had to get rid of. This was turning out to be a costly process and we desperately needed money for more urgent things.

Life at last was taking on some sort of order. Gradually we were acquiring more implements and labour-saving devices, although we still also had a lot of bailer twine and old wire makeshift devices too, but we were just a bit more in control of our destiny. We had learned so much in the first three years but the progress had been slow and painful. We now had veterinary books on most of the common cattle ailments and could apply a lot of remedies ourselves, without necessarily relying on the vet all the time. We could spot tell-tale symptoms of illness early on and we had become generally more efficient. Farming is a hard life at the best of times and the rewards can sometimes be few. I remember feeling happy just because something hadn't gone wrong, let alone the bonus of something going profitably right!

Little by little our pioneering was starting to show some signs of working. We got the odd cheque from sales at market usually, it is true, only to get it swallowed up again when the feed costs dramatically escalated because of oil prices soaring. Then the cost of manufacturing cattle fodder went through the roof. All this had to be added to the long eighteen months of rearing calves, along with the vet's bills, inoculation serum and endless other expenses, before you could ever show a profit. Then one day, you send a batch of bullocks to market and if it happens to be a bad day and beef prices are down, you are lucky to have cleared your overheads. Of course you can always take them back for a few months but that is double transport costs and we were far too small to be able to afford a cattle truck of our own. There were no guarantees even then, that if you returned them to market later, that it wouldn't be a bad day again.

The farm was subsidised totally with Conrad's acting and the odd repeat fee that floated in from time to time. With two small children, I was in no position to go out to work. Wages in Dumfries were not particularly good, even on the secretarial side, and besides I was putting my physical whack into the farm, alongside Conrad. It needed both of us full time. But we were making progress. It was an all-absorbing lifestyle and we were quite self-sufficient in many ways. What was really happening in the outside world seemed a million miles away and it only intruded into our lives through the odd television programme or letter. Most of our time was taken up with our own basic survival.

One of the most coveted achievements, for our farming neighbours, was to have a mention in The Scotsman, Scotland's leading newspaper, for obtaining the best price at market. The paper held a lot of clout and rivalry in the farming community - to get the highest price, was rife. Our neighbour had been trying all his life, with no success, in spite of the fact that his was an extremely well-mechanised, efficient farm of some several hundred acres. At the time we were oblivious to this piece of information and, I think, totally unaware of The Scotsman's existence.

As the months passed, Juno got fatter and fatter and sleeker and still unproductive. Reluctantly Conrad and I decided she would have to be sold on. Maybe some other farmer would have better luck. Maybe she would be better off taken to the bull. Either way, she was not paying for her keep and so one spring morning, she was taken to Dumfries market. We were sorry to see her go. She had given a lot of joy but we were desperately trying to get more efficient and in this instance, the head has to rule the heart.

We arrived at the market shortly after she had been unloaded and took place a few seats up from the ring side, where we had a better view. Business was brisk and some fairly good prices were being achieved. We were tense with anticipation. The auctioneer was doing his usual fast delivery and there was a buzz of energy in the place. Suddenly in came Juno, looking very wild eyed but also very sleek and well cared for. The bidding started at a fast and furious pace. Juno charged around the ring, glaring balefully at all. The bidding continued for a long time. We couldn't keep up with it. Finally the gavel went down and that was it. We had no idea what she had fetched until the end of the sale and we queued for our cheque. We were very pleased. It seemed too good to be true and we immediately went off to the Castle Arms to celebrate.

Some weeks later, after all the excitement had died down, there was a knock at the door. Our neighbour was standing there, his cap in his hand. The usual dew drop was frozen to the end of his nose. He was holding a sack in the other hand.

"Mornin' missus - 'sa grand morning, aye." Yes I agreed, it was only minus 5 degrees - lovely.

"I brought ye a few neeps (turnips). Thought ye might like 'em".

I felt sure there was something behind this sudden burst of generosity.

"Is ya man aboot?" he queried.

"Yes," I said. "Come in".

"I'll no stop long," he said. I called out to Conrad. The old man greeted him. Now praise was not something that came easily to him but his blue eyes twinkled. Rubbing his nose thoughtfully for a few moments, he gazed heavenward and gave a sigh.

"I just came to congratulate ye".

"On what?" queried Conrad.

"The mention. Ye got the mention I see ...in The Scotsman, last week. Best price. Never achieved it mesel - yet. But anyway - congratulations". He reflected, clearly mystified... (English, townies too!) With that he gave a curt nod, put his cap back on and paced out of the yard. Wow! Praise from him was praise indeed.

I put the kettle on for coffee and we sat down and had a little laugh. A mention in The Scotsman. A lifetime's ambition, achieved by a pair of townies. We started to reflect ourselves. Life in the Fulham Road was a million miles away now. What bravery. What stupidity. We had made so many mistakes and doubtless there were those who viewed us patronisingly, sure that we would fall at the first fence. After all, what did we think we were doing, an actor from London, playing at farming but although we fell over many fences, when we picked ourselves up and viewed our lives, were the alternatives so blooming marvellous?

Suddenly it all seemed worth it. All the trials and tribulations and tears. We had pitted ourselves against the elements, with enormous odds against us. We had started out knowing nothing at all about farming. Had total despair, not even average luck and very little money, but we had learned. Here we were - nearly four years on and for goodness sake - a mention in The Scotsman!

ABOUT THE AUTHOR

Jennie Phillips lived her early life in London before moving to Skeoch and then emigrated from there to Northern France where she and her actor husband, Conrad, set about restoring a dilapidated farmhouse. This was all despite neither of them speaking French. After 20 years spent in France, she now lives in Wiltshire where, apart from writing, she also sells paintings (mostly landscapes), plays music, knits (including copious numbers of sweaters for her grandchildren) and cooks.

Among Jennie's favourite authors are Maureen Lipman, Derek Tangey and Ken Follett and she particularly likes reading about the emotional lives of people and poetry.

Her particular and heartfelt thanks go out to her family and her husband, Conrad. for all their support and encouragement towards writing Skeoch.

Jennie's blog - jenniephillipsblog.wordpress.com

Printed in Great Britain
by Amazon.co.uk, Ltd.,
Marston Gate.